Subcontracting E

A Management and Te(

Subcontracting Electronics
A Management and Technical Guide for Purchasers and Suppliers

David Boswell

McGRAW-HILL Book Company

London · New York · St Louis · San Francisco · Auckland · Bogotá
Caracas · Hamburg · Lisbon · Madrid · Mexico · Milan · Montreal
New Delhi · Panama · Paris · San Juan · São Paulo · Singapore · Sydney
Tokyo · Toronto

Published by
McGRAW-HILL Book Company Europe
Shoppenhangers Road · Maidenhead · Berkshire · SL6 2QL · England
Tel 0628 23432; Fax 0628 770224

British Library Cataloguing in Publication Data
Boswell, David
　　Subcontracting Electronics: A management and technical guide for purchasers and suppliers
　　I. Title
　　621.381'068

　　ISBN 0-07-707783-0

Library of Congress Cataloging-in-Publication Data
Boswell, David
　　Subcontracting electronics: a management and technical guide for purchasers and suppliers/David Boswell.
　　　　p.　　cm.
　　Includes bibliographical references and index.
　　ISBN 0-07-707783-0
　　1. Electronic industries—Subcontracting.　2. Production management.
I. Title.
TK7836.B67　1993
621.381'068'7—dc20　　　　　　　　　　　　　　　　　　　　　　92-31693
　　　　　　　　　　　　　　　　　　　　　　　　　　　　　　　　　CIP

Copyright © 1993 David Boswell. All rights reserved. No part of this publication may be reproduced, stored in a retrieval system, or transmitted, in any form or by any means, electronic, mechanical, photocopying, recording, or otherwise, without the prior permission of David Boswell and McGraw-Hill International (UK) Limited.

12345　CUP　96543

Typeset by MFK Typesetting Ltd, Hitchin, Herts.
and printed and bound in Great Britain at the University Press, Cambridge.

Contents

Foreword	xi
Introduction	xiii

1 The electronics industry	**1**
1.1 Industry structure	1
1.2 World electronic industry subcontract market size	3
1.3 Types of assembly subcontractor	4
1.4 Component procurement options	7
1.5 Test house subcontractors	7
2 Key factors in surface mount 'make or buy' decisions	**9**
2.1 Making in-house: the benefits	10
2.2 Subcontracting: the benefits	13
2.3 Making in-house: budgeting for the costs	14
2.4 Making in-house: a disadvantage	22
2.5 Subcontracting assembly: budgeting the costs	23
2.6 Subcontracting: some disadvantages	23
2.7 Conclusions	24
3 Contracts, terms and conditions of purchase and sale	**26**
3.1 Types of contract for board assembly	26
3.2 Terms and conditions of sale and purchase	28
3.3 'Turnkey' projects	30
3.4 Arbitration	33
Appendices	
3A Example agreement terms between an OEM and a subcontractor	34
3B Typical subcontractor's Terms and Conditions of Sale	42
4 Assembly technology options, advantages and disadvantages	**48**
4.1 Introduction	48
4.2 Through-hole (conventional) printed board assembly	48
4.3 Surface mount printed board assembly	52
4.4 Mixed-technology printed board assembly	57
4.5 Hybrid (film) circuits	59

4.6 Tape automated bonding (TAB)	62
4.7 'Chip on board' assemblies	64
4.8 Multichip modules (MCMs)	66

5 Design checks before subcontracting SM assemblies — 69
5.1 Background comments	69
5.2 Electronic circuit design	70
5.3 Component selection	71
5.4 Printed board specification and layout	71
5.5 Large printed board assemblies	73
5.6 Equipment structure	74
5.7 Product safety	75

6 Design checks before subcontracting hybrid circuits — 79
6.1 Background comments	79
6.2 Electronic circuit design constraints	79
6.3 Substrate layout and structural constraints	80

7 Surface mounted components and assembly materials — 83
7.1 Solderability of components	83
7.2 Specifying and purchasing surface mounted semiconductors	85
7.3 Specifying and purchasing typical passive components	86
7.4 Goods inwards inspection, storage, and kitting of surface mounted components for high yield in assembly	88
7.5 Adhesives and solder pastes	90

8 Printed boards — 92
8.1 Specifying and purchasing surface mount printed boards: a checklist	92
8.2 Bow and twist in large printed boards	102
8.3 Goods inwards inspection and storage of SM printed boards for high yield in assembly	103
8.4 Long-term storage and baking of printed boards	105

9 Components supplied by the customer — 106
9.1 Free-issue scenarios	106
9.2 The customer sells components to the subcontractor	109

10 Checking designs for manufacturability, test and reliability — 110
10.1 Introduction	110
10.2 Designing for surface mount technology	110
10.3 Design for manufacture	111
10.4 Checking the layout	115

11	**Specifying the product and defining the basis for quote**	**119**
	11.1 Introduction	119
	11.2 Defining the product	119
12	**Choosing a printed board layout design subcontractor**	**128**
	12.1 The designer	128
	12.2 The CAD system	129
	12.3 The job to be done	130
	12.4 Conclusion	133
13	**Choosing an assembly subcontractor**	**134**
	13.1 Financial aspects	134
	13.2 The subcontractor's team	135
	13.3 Defining the capability and capacity requirements	135
	13.4 Questions asked by professional assembly subcontractors	137
	13.5 Reviewing quotations	137
	13.6 Sample quotations	138
	Appendices	
	13A Example quotation for prototype assembly	139
	13B Example quotation for production quantities	145
14	**Choosing a customer: the subcontractor's viewpoint**	**152**
	14.1 Financial and market aspects	152
	14.2 Technical aspects	154
	14.3 Status of the customer's structural design	155
	14.4 Status of the customer's electronic design	156
	14.5 Quality aspects	156
	14.6 Design for manufacture and test	157
15	**Assessing and auditing a subcontractor's assembly and test activity**	**158**
	15.1 Control of materials	158
	15.2 Process control	161
	15.3 The quality function	164
	15.4 Visual standards, inspection and rework procedures	166
	15.5 Electrical testing	169
	15.6 Traceability requirements	172
	15.7 Quality assurance testing	173
	15.8 Anti-static precautions	174
	15.9 General housekeeping and discipline	176
	15.10 Trouble-shooting capability	177

16 Rework and repair — 179
16.1 Background — 179
16.2 Assessing rework equipment and processes — 180
16.3 Key problems in surface mount rework — 181
16.4 Selection of suitable rework equipment — 184
16.5 Preparation for rework and repair — 187
16.6 Rework activity classification — 188
16.7 Rework machines, tools and methods — 191
16.8 Rework recording procedures — 203
16.9 Field repair philosophy — 204

17 Surface mount field reliability issues — 205
17.1 Introduction — 205
17.2 Multilayer ceramic chip capacitor failures — 206
17.3 Failures in large semiconductor integrated circuit plastic packages — 212
17.4 TCE mismatch — 213
17.5 Other reliability pitfalls — 214
17.6 Conclusion — 214

18 Maintaining good relations — 215
18.1 Defining respective responsibilities — 215
18.2 Departmental contacts — 215
18.3 Controlling changes — 216
18.4 Customer acceptance criteria and return procedures for large or complex boards — 218
18.5 Product safety and product liability issues — 221

19 Costing board assemblies — 223
19.1 Background comments — 223
19.2 Labour-based cost-estimating methods — 225
19.3 Machine-based cost-estimating methods — 226
19.4 'Scarce resource' costing — 226
19.5 Mixed machine and labour-based costing — 229
19.6 Typical labour-based subcontractor cost estimate — 229
19.7 Tooling costs — 232

20 Compound case-histories	**233**
20.1 Dataphlog Ltd: an in-house production disaster	233
20.2 Buckpass Ltd: a subcontract tragedy of errors	242
20.3 Rufflay Ltd: a CAD bureau story	246
21 Definitions and abbreviations	**252**
22 Sources of assistance and bibliography	**259**
Index	**260**

Foreword

The make-or-buy question is being asked increasingly in the electronic business. This book comprehensively addresses all the key drivers to be considered in arriving at a decision, and also examines the key issues that need to be understood and controlled if the buy decision is to be successful.

Contract manufacturing is growing in response to continuing changes in technology, and there is a need for more flexible manufacturing and continuous cost reduction programmes.

In this book, David Boswell explains the need for partnership to ensure that contracting works profitably for both parties. It should be used as a reference by all senior executives considering contract manufacturing, and is written so that they can explore those issues that might affect their particular circumstances.

Partnership is a word that is easily used but takes effort, commitment and resource from both parties. Contract manufacturing is a partnership business.

David Pratt
Managing Director, Philips Circuit Assemblies
Chairman, Association of Contract Electronic Manufacturers

Introduction

The guidance offered in this book is intended to help all who work in the electronics industry subcontract arena, whether they be purchasers or suppliers. It takes the reader, step by step, through the stages prior to, during and after contract placement. It focuses on the pitfalls and how to evade them. It sets out the necessary high standards of management for both customers and subcontractors, especially where new and mechanized assembly methods are used.

Although surface mounting is used to illustrate several of the main messages, much of the content of the book is applicable to all forms of electronic construction.

Interconnection technologies are becoming progressively more complex and require a far wider range of knowledge in their application than in the past. In parallel with this trend, we see that the increased number of variables in design and manufacture of miniaturized circuitry is demanding a greater degree of professionalism in controlling them.

Many of the elements in the 'make or buy' decision, for example, in surface mount assembly suggest a steady move towards more subcontracting by original equipment manufacturers (OEMs). Aside from the major capital expenditure needed for in-house production of reliable products, the cost of the learning curve is usually higher than forecast and, by its nature, may not be obvious to management until it is too late. The importance of affording instant technologist support on the assembly line is another area where lack of awareness at high level has sometimes led to excess costs.

Subcontracting avoids many, but not all, of these problems. It certainly brings different ones to the fore. But the key attributes of attention to detail and planning for successful subcontracting are not very different from those needed for in-house production. If the quality of management is inadequate in the subcontract scene, however, the risk of misunderstanding, delay and loss of profit for both parties is disproportionately greater.

To optimize 'design for manufacture' and achieve minimum cost, the customer nowadays should select a subcontractor at an early stage and be aware of specific machine limitations while designing the electronic circuit, and certainly well before laying out the printed circuit board. The old idea of leaving it to the purchasing and manufacturing departments to sort out basic production methods after all the design work is complete is a recipe, if not for disaster, then at least for high cost and sometimes for reliability problems.

In this book, the sequence of events preceding and following order placement are looked at from both the customer's and the subcontractor's point of view. Typical situations and methods of assessing subcontractor capability are discussed, including the use of checklists containing questions that both parties should raise and answer.

Procurement of surface mount components today demands more stringent definition of the required goods than in times past. The printed board layout is based on terminations having strict geometric disciplines. A part that is electrically equivalent but delivered in the wrong package can no longer be adapted to fit. Here is one freedom of manoeuvre that has been sacrificed to gain further miniaturization, higher speed and lower assembly cost.

Also, there are now several alternative mass soldering methods available. The electronic design and printed board layout should be carried out with full knowledge not only of the intended assembly processes, but also of the ability of the specified parts to survive them. Before placing orders with subcontractors, wise purchasing managers will protect their own and their companies' interests by seeking confirmation that this has occurred.

The means by which the subcontractor receives materials for the product has always been crucial to a good relationship, but the advent of reflow soldering has placed more emphasis on the quality and age of surface mountable components than on just-in-time (JIT) practice. Normally it is the components that form the largest element in assembled board cost. For this reason, the pros and cons of 'free issue' and the heightened importance of goods inwards inspection demand careful consideration.

At the heart of all subcontract operations is the management of change. The way this is handled often marks the difference between the professional and the amateur subcontractor. Speed of response is key, but there is no substitute for the written word!

While there is evidence emerging that well-made surface mount components and assemblies are no less reliable than conventional through-hole structures,[1] a new set of sensitivities must be considered. These arise from the higher temperatures seen by components during soldering and the comparative rigidity of their attachment to the printed board. In setting up a subcontract project employing any new assembly technology, it is vital that key reliability aspects do not disappear in a gap between the two parties.

To ensure that the right messages reach those readers who are scanning the chapters with a limited field of view, there is a small element of repetition

---

1. 'Surface Mount Technology: The message from the field', J.A. Hayes, Proc. Smartex Conference, Wembley, October 1992.

between some sections of the book. No apology is made for highlighting and focusing on many of the difficult technical areas in surface mount technology, for example rework, reliability and related subcontracting aspects. For executives with restricted time to read books, information on pitfall avoidance is among the most important aspects to cover.

Experience has shown that the financial penalties for failing to meet these new challenges in a professional way can be very severe. Both parties need to accept revised behaviour patterns if their relationships are to be compatible with today's needs in the electronics industry.

The 'duty of care' towards the end-user is paramount.

Acknowledgements

I am most grateful to Martin Wickham of the UK Surface Mount Club for the initial preparation of Chapter 16 on Rework and Repair and for his excellent support in proofreading. My thanks also to Peter Nolan of Surface Electronics for permission to adapt the Terms and Conditions of Sale and the product safety guidelines I originally prepared for them and for the further advice thereon.

The McGraw-Hill Book Company have kindly agreed to my incorporating suitable extracts from their *Surface Mount Guidelines for Process Control, Quality and Reliability*, by David Boswell and Martin Wickham (1992). These extracts occur mainly in Chapters 7, 8, 10 and 16.

Important notes

This book is for guidance only. Users should check that their materials, processes and products meet required professional standards and/or, where appropriate, national and international standards.

Neither the author nor the publisher accepts liability for any action or for the consequences of any such action taken by the user based on the guidance given in this book.

By courtesy of David Boswell and Technical Reference Publications Ltd, some extracts from the following volume have been incorporated in this book (these are identified in the text):

> *Surface Mount and Mixed Technology PCB Design Guidelines* by David Boswell (1990), Technical Reference Publications, Asahi House, 10 Church Rd, Port Erin, Isle of Man, UK.

1

The electronics industry

1.1 Industry structure

The emergence of the electronics industry as a separate entity began with the development and widespread sale of radio sets in the second quarter of the twentieth century.

In the very early days, apart from the basic Galena crystal, the thermionic valve was the key component, and those that made the valves also made the sets. To begin with they had to manufacture their own baseboards, valveholders, resistors, capacitors, coils, and often the polished mahogany cabinets as well.

Thus, the electronics traders developed as a group of vertically integrated organizations struggling to achieve recognition as a force separate from the rapidly growing electrical business sector. Most were spin-offs from major companies who were already involved in power generation and in the manufacture of electrically driven equipment for homes and factories, but there were also many small start-up firms driven by enthusiastic entrepreneurs.

The above industrial scenario in the UK and the USA was reflected in the politicking that took place within the learned institutions and trade associations that had grown up around the 'DC to 50 cycles' business. The 'electricals' tried hard to absorb the 'electronics' tide and to retain their dominance, for example, in the affairs of the UK's Institution of Electrical Engineers (IEE). In the latter case, such tactics led eventually to the breakaway formation of the rival Institute of Electronic and Radio Engineers (IERE) and others. The IEE was slow to accept electronics as a vital element in its deliberations and, above all, in its organizational hierachy. Today electronics is the dominant group in that august body.

As 'know-how' proliferated and competitive forces emerged, the industry structure changed. The divide between component manufacturers and setmakes was born, and later the providers of raw materials became a third, albeit scattered, force in the electronics arena.

The same general development pattern was seen in the early stages of the emergence of discrete semiconductor devices. The very high investment needed meant that they rapidly became the prerogative of the giant equipment companies like Bell and IBM, but this time the existence of an already powerful components industry quickly caused a reverse trend to occur.

With the advent of silicon integrated circuits, several independent semiconductor companies, thirsty for added value and realizing the strength of their technological position, began to make and sell a wide range of products containing their components, e.g. from watches to computers and test equipment. Some even tried to battle head-on with their best customers—they lured away their most experienced staff and then offered competitive products.

It was the advent of the Second World War that gave rise to subcontracting assembly in the UK and US electronics industries. Certainly in Britain it became necessary to take the work stemming from the rapid expansion of military electronics to workforces scattered out of range of enemy bombers. The government offered incentives (e.g. the 'cost plus 10%' scheme) to many small entrepreneurs in mechanical assembly to provide this service. It worked.

Subcontract assembly has since steadily increased its importance worldwide. When labour-intensive work was necessary, it became fashionable for larger US and European companies to seek low-labour-cost areas in the Far East. The semiconductor industry began this, but was soon followed by OEMs seeking to avoid the high protective import tariffs imposed by Western European governments on semiconductor imports. The intention was to preserve their indigenous key component (meaning semiconductor) industries, but the effect was to drive the larger OEMs to source both semiconductors and equipments overseas.

This description is simplistic, but is intended to highlight the change brought about by reduced tariffs and highly mechanized assembly such as we now see in surface mounting. The criteria for locating manufacture with minimum cost focus no longer on cheap labour, but on the combination of best infrastructure at technician level and local government capital investment grants. This has encouraged the return to Europe of some major computer product lines and has also benefited Far Eastern companies wishing to seek European production bases before the Common Market becomes a reality.

Noteworthy, too, is the European swing from protecting indigenous semiconductor manufacture to protecting electronic systems manufacture. The validity of the initial decision was open to doubt, but, having lost the battle, European firms now appear to be seeking salvation at the next level up in the constructional hierarchy. Success will probably be denied them by the massive inward investment invited to offset the electronic and other industries' unemployment problems—many of which arose from the original errors of judgement on protectionism. This comment applies to the combined efforts of successive governments—it is apolitical.

In the current situation, the dividing lines between component and equipment makers have become progressively more blurred. If we add the assembly subcontractors and makers of modular parts to the equation, today's industry structure has become even less definable.

Perhaps the best way to describe things is to consider the present scene as a flexible, see-through partitioning set-up, with designers plus marketeers on one side and 'nuts and bolts' merchants on the other. In the larger companies the divisional structures are tending to become less vertically integrated, and in following this trend we are seeing a reversal of the compact, product-oriented cost centre arrangements favoured in the 1970s and 1980s.

The tasks of today's executives in electronics are made more demanding by the increasing flexibility of our industrial hierarchy and the returned risk of 'eddy current management'—a jocular reference to the inefficient dual-line reporting systems used by some major world conglomerates in the 1970s. The harsh fact is that any further separation between design and manufacturing functions must be regarded as a retrograde step for the electronics industry as a whole and in particular for the successful practice of 'interconnection technology'—the hardware IT that makes the software IT (information technology) possible!

It is this type of segregation problem that has brought the emergence of new pressure groups within the electronics industries of the developed nations. One such body in the UK is ITIC, the Interconnection Technology Industrial Consortium. Its remits are twofold: one is to make the right government bodies aware that pouring comparatively large proportions of the national R&D funding budget into esoteric semiconductor devices will be unwise unless the right interconnection technologies to take advantage of them are also developed; the other is to make universities, polytechnics and technical colleges more responsive to industry's needs by including interconnection technologies in the syllabuses for electronics degrees and similar qualifications.

Designers and engineers in this field who sit before CAD (computer-aided design) machines now need much more than a basic grounding in physics, chemistry, plastics and mechanical engineering to augment their knowledge of electronics. The provision of suitable multi-discipline curricula is vital to the 'design for manufacture' concept.

1.2 World electronic industry subcontract market size

Assessment of market size for the subcontracted portion of electronic assembly in developed countries has been difficult because of a lack of data collection. Until the formation of organizations representing the majority of subcontractors occurs, at least in Japan, Europe and the USA, spasmodic superficial forays by market analysts will have to remain the only substitutes.

Today there are many very small board assembly outfits all over the world, typically with 10 or fewer people operating on what should be described as a 'cottage industry' basis. They provide a useful satellite service to nearby OEMs, but are not included in subcontract business statistics because they cannot afford the membership fees demanded by most trade association organizations.

It was thought originally that the advent of surface mount technology would diminish the number of small companies in this category because of the high capital investment required for reliable processing. In the event, the reverse seems to be true, since, in addition to those diversifying, a number of new facilities dedicated to surface mounting and spurred on by the emergence of cheap, low-volume processing equipment have sprung to life (cf. Section 1.3, Class 4).

The proportions of board assembly work subcontracted vary widely between countries. Compared with the USA, (West) Germany and France, the amount put outside by UK OEMs is large—probably as high as 22 per cent compared against 10–15 per cent in those countries.

The subcontracted portions of board assembly vary even more widely between industry sectors. In automotive electronics the percentage can be as high as 70 per cent, while in defence electronics it is usually around 10 per cent.

Tables 1.1 and 1.2 represent my own methodology in assessing the order of magnitude of subcontract market size across the various industry sectors and on a worldwide basis. From the figures given in these tables, one may estimate a 1992 total subcontract activity within Europe of the order of $8 billion, of which the UK contribution would be around $2 billion. In arriving at these figures, it should be borne in mind that:

1. Many equipment products shipped from European OEMs contain boards assembled offshore—mainly in the Far East.
2. The figures exclude 'turnkey' projects in which the subcontractor makes the entire equipment.

1.3 Types of assembly subcontractor

Independent assembly subcontractors can be classified into four main types.

Class 1
These companies usually have sales in excess of $20 million and offer a complete design, manufacture and testing service, including product assessment using temperature cycling and other methods. These are the 'majors'

in the field, and their engineering departments are large enough to have expertise in several areas of electronic circuit design as well as in the board CAD layout and mechanical design of assemblies and complete products. They are therefore capable of handling major projects on a 'turnkey' basis.

Some of these subcontractors will set a lower limit on the magnitude of work they are prepared to undertake, typically $100 000 p.a. per product.

Table 1.1 Total world output of electronic equipment, 1990 ($billion)

		EDP	Office	Telecom.	Cons'r.	Auto.[†]	Indust.	Med.	Defnce	Total
1	USA	58.0	5.7	16.8	5.8	19.0	27.2	5.8	45.4	184
2	Japan	63.2	5.4	15.1	33.9	18.0	9.2	3.9	6.1	155
3	Europe	16.4	3.9	25.0	12.5	9.0	21.5	4.0	7.2	100
4	Rest	22.2	1.7	9.3	19.8	4.0	3.4	0.8	5.7	67
5	Totals	159.8	16.7	66.2	72.0	50.0	61.3	14.5	64.4	506

Table 1.2 Derivation of total world subcontract market size, 1990 ($billion)

		EDP	Office	Telecom.	Cons'r.	Auto.[†]	Indust.	Med.	Defnce	Total
6	Prime cost[*]	78	8	30	42	28	31	6	30	253
7	Board ass'y[*]	47	3	17	13	20	12	2	5	119
8	Subcontr't[*]	12.0	0.5	4.5	2.5	10	2.4	0.4	0.5	33

[*] Includes value of components and printed boards.
[†] Figures unsubstantiated.

Key
EDP   Electronic data processing equipment, computers
Office   Office equipment
Telecom. Telecommunications equipment
Cons'r.   Consumer, brown and white goods, toys
Auto.   Automotive equipment
Indust.   Industrial controls and instrumentation equipment
Med.   Medical equipment, passive and active
Defnce   Defence, avionics and space equipment

Notes
Line 6 is based on assumptions of typical gross margins in the various industry sectors.
Line 7 is based on assumptions as to the percentage of prime factory cost represented by the board assembly, including components and boards.
Line 8 assumes typical subcontract percentages appropriate to industry sectors.

Sources: BISMackintosh, Surface Mount Intelligence Group, ECIF (UK).

Class 2

With turnovers ranging from $5 million to $20 million, in addition to assembly, these companies offer a board layout service via in-house or subcontracted CAD facilities and a limited degree of electrical test capability. Their interests will lie in project sizes over $50 000 p.a., and they have sufficient engineering support to give strong inputs to customers on board layout, process choice and technical problems in manufacture.

Class 3

This class of subcontractor operates in what may be described as the 'passive' mode. Such companies tend to be purely assemblers and offer little or no comment on their customers' board designs prior to order acceptance and manufacture. They take an optimistic view on these, and their turnover size range can be large because of the attitude adopted, e.g. between $20 million and $200 million. This means that they can operate with low overheads, but in consequence they are saddled more often with the burden of customers' poor designs.

Class 4

These are the small companies—also with low overheads, sometimes unkindly described as 'garage operations'. Their turnover usually ranges between $0.2 million and $1 million and they are often sited close to large OEM customers. They also tend to use the 'passive' approach and will specialize in prototype and small-scale production orders. Many will have diversified from conventional through-hole assembly services and may not necessarily have the equipment or the expertise to control all aspects of surface mount quality and reliability problems.

Overlaying the classification into four groups, there is another category of subcontract service available from time to time. This comes from OEMs who have set up in-house facilities for their own needs but have temporary spare capacity which they offer to the market. These capabilities lie mainly in Class 3.

However, some potential customers are wary of placing important or long-term work with the OEMs, believing that if there is a sudden upsurge in in-house production requirements their projects will go to the back of the queue. There have in fact been several cases where one month the OEM was touting for work and the next month it was seeking spare capacity among its independent colleagues. The latter, whose living is based solely on subcontract business, get upset when they lose work to one of these operations, especially if the order has been taken on a marginal cost basis, i.e. just to recover overheads. Nevertheless, these operations can provide a useful low-cost service to other OEMs having urgent short-term needs.

Memo: It is known that much of the small, low-cost soldering equipment sold in the market-place has not always been suitable for producing reliable assemblies with all component types. OEMs are strongly advised to proceed with caution when seeking subcontract services from assemblers having this type of equipment. Chapter 17 explains the problem.

1.4 Component procurement options

Normally, in the first three classes, the options of working with free-issue parts or on the basis of total (or partial) procurement by the subcontractor are available. Subcontractors in Class 4 are more likely to prefer free-issue supplies. Either way, there are many possible contract variations between the complete product design and manufacture (turnkey) situation and a straightforward 'Here are the bits, get on with the assembly' relationship.

While some subcontractors are known to dislike the degree of control exercised by their customers and the irritation of tripartite quality negotiations which can sometimes characterize the free-issue component scenario, they may nevertheless prefer these risks to the potential heavy cash flow commitment inherent in a total procurement service.

In practice, most assembly subcontractors try to maintain a balanced mix of project material supply options. This subject is discussed in greater depth in Chapter 9.

1.5 Test house subcontractors

Many small OEMs and assembly subcontractors are not equipped with automatic test equipment (ATE) capable of handling large complex circuit boards which need diagnostic capability and calculation of sophisticated statistical performance data. Such equipment can cost upwards of $500 000 and requires both a high standard of programming knowledge and skilled maintenance.

A solution is to make use of specialist test houses which have invested in extensive test facilities of this type and provide a service to industry. Some have several test centres in a given country so that they can offer their capabilities on a more localized basis. In addition to the ATE and appropriate software specific to the customer's circuitry, they also provide bespoke hardware, e.g. test rigs and probe systems.

The obvious disadvantage in using test houses is the time lost when assemblies are in transit to and fro. To overcome this difficulty, some offer to place one of their equipments within the OEM or assembly subcontractor's premises and to provide a managed on-site service. A deal of this type would include a regular rental charge independent of the load provided by the customer, plus unit costs based on test times for individual circuits. Typically, there might also be a rapidly escalating scale of charges for retesting reworked circuits; and the test house often retains the right to use the equipment for its own purposes when it is not required by the customer. This latter part of the arrangement is usually offset by the test house agreeing to provide alternative facilities on its own site when the bespoke equipment cannot cope technically with the requirement—for example, when it has insufficient test node capability or is not fast enough.

2

Key factors in surface mount 'make or buy' decisions

Competition and smaller market opportunity windows are demanding faster design and manufacturing responses within the electronics industry.

Electronic equipment life-cycles are steadily shortening. In consequence, 'start-up' and 'phase-out' periods are assuming far greater proportions than hitherto. This phenomenon is of vital importance in assessing the profitability of investment in new product development and in the technologies employed in their manufacture.

At the same time, 'make or buy' decisions have become progressively more complex. Cost–benefit analysis has to embrace a wider range of factors, and their quantification may frequently prove impracticable within available knowledge, budgets and time-scales.

For this reason, the debate in smaller companies will often focus more on philosophical aspects than on the 'nitty-gritty' of financial comparison. There is a viewpoint that this is no bad thing because, time and again, we see that the best business decisions turn out to be those made against the quantified odds—and vice versa.

A corollary of the above statements is that it is now clear that the 'make or buy' issues in today's assembly technologies must be addressed not only by experienced high-calibre staff, but also at the highest possible level of management—i.e. at Board level. This chapter seeks to outline the major agenda items that should be considered in looking at surface mounting, and Chapter 20 illustrates the seriousness of some of the pitfalls in 'case-history' form.

Before going further, we must acknowledge the importance of internal politics within a company. The production manager who favours subcontracting is likely to be a rare bird, and top executives need to be fully aware of the enthusiasm that may permeate any proposal coming from that quarter. The problem is that often it is the only low-cost source of planning information specific to the company's needs.

Usually it is the managing directors (MDs) who will benefit their companies most by receiving training in surface mounting before taking the decision, but, in my experience, frequently they will cling to the mistaken belief that they are the people least in need of it. Ideally an expert third-party opinion should be sought when there is doubt, or when it is clear that the management team has had little experience in the new technology involved.

Some say that you can teach accountancy to an engineer but you cannot teach engineering to an accountant. This proverb may carry a grain of truth, but the understanding of all the factors that should be considered in today's 'make or buy' electronic assembly decision is just as likely to be adversely affected by a lack of accounting and/or technical awareness in a managing director as by a non-scientific approach from a financial wizard.

Some useful items for inclusion in a cost–benefit analysis covering surface mount production capability are offered below. They are not intended to be exhaustive, but it is hoped that the condensed checklist given in Table 2.1 may at least alert management to some of the significant, but less well publicized, problem and benefit areas, particularly in M6 and M7 below.

Discussion on the direct comparison of manufacturing cost in-house versus the purchase price from an assembly subcontractor has been avoided deliberately. The balance of results can vary as each individual circuit is considered. In any event, experience shows that the projected figures in a comparatively new technology are likely to be markedly different from reality and, as indicated in the first case-history in Chapter 20, can be somewhat dependent on the motivation of their compiler.

Table 2.1 summarizes the subjects reviewed in this chapter, and the 'M' and 'S' subheadings follow the nomenclature used in the table.

2.1 Making in-house: the benefits

M1 Marketing tool
A successful mechanized surface mount assembly line in full production is an impressive sight. However, in my experience of important customer visits, either all the robotic 'pick-and-place' machines are exercising their uncanny ability to select such days for complete malfunction, or they are being re-programmed at the time of the factory tour.

Some companies have deliberately moved early into surface mount technology for commercial rather than technical reasons. For example, a well-known maker of navigation and radar equipment for small boats took the plunge in 1984, although there was no serious size problem for them. Their competitors felt it necessary to follow suit.

Three years ago companies offering products containing surface mounted

Table 2.1 Surface mount assembly: 'make-or-buy' cost–benefit considerations

M:	Make in-house	S:	Use subcontractor
	Benefits		
M1	Marketing tool	S1	Spreading financial risk
M2	Vertical integration	S2	Choice of technology
M3	Fast response/priority control	S3	Reduced technical risk
M4	Commercial security	S4	Choice of supplier
M5	Knock-on effects		
M6	Reorganization option		
	Budgeting the costs		
M7	Management training	S5	Full specification
M8	Capital investment	S6	Second sourcing
M9	Space and services	S7	Production runs
M10	Technologist support	S8	Free issuing components
M11	Design rules and training	S9	Loss of added value
M12	CAD equipment		
M13	CADMAT		
M14	Component sourcing		
M15	Production control		
M16	Operator training		
M17	Start-up losses		
M18	Staying in the game		
M19	Exit phase		
	Disadvantages		
M20	Technology stretch syndrome	S10	Late delivery risk
		S11	Testing

components were claiming the use of the new assembly technology as a bonus point in their sales advertising. Today the world accepts surface mounting as the norm in many applications, and it is not considered worthy of mention except when advanced versions such as tape automated bonding (TAB) are included in the construction (cf. Section 4.5).

M2 Vertical integration

'Design for manufacture' is seen as a developed country's pathway to competitiveness on a worldwide basis. There is little doubt that vertical integration can be a key element in success on this score, but, unfortunately, many financial managers in the electronics industry are currently seeking greater short-term benefits in 'rationalization'. This word has become a euphemism for pooling design and production resources in physically separated centralized groups and then failing to provide the management and communication infrastructure necessary for their medium- and long-term survival.

At its optimum, this approach is a second-best solution, as many large European OEMs who placed their equipment assembly lines in the Far East five years ago have discovered. Several are busy bringing them back—and not only for European Community tariff barrier reasons. Nor does it take account of the antagonisms that often arise between separated divisional profit centres within the same company or group. These centres do not like to do business with each other because they know that the other party can easily call upon 'head office' for support. When head office decides to administer 'help' to the supplying party, this usually arrives on their doorstep in the form of a senior executive carrying a big stick and having the power to change their priorities and reduce their profit for the sake of their sister division and the company as a whole. The MD of the 'helped' division is later castigated for failing to achieve his output and profit targets!

In-house capability and first-hand knowledge of the practical aspects within the company may be essential in obtaining some types of OEM system contract, particularly where both design and manufacture are involved for military projects.

M3   Fast response time and priority control
These should prove helpful and easier to manage in-house than with subcontractors, though obviously this will depend to some extent on workloads.

M4   Commercial security
It is argued that in-house manufacture provides bettter commercial security than subcontracting, but hard evidence is lacking. In any case, the possibility of leaks via key people leaving the company can never be entirely eliminated.

M5   Knock-on effects
Undoubtedly, in-house activity in a new technology can spark off ideas for new products and methods.

M6   Reorganization option
The reorganization needed when setting up the new technology can be used as a vehicle by management for implementing other necessary reforms, e.g. 'just-in-time' (JIT) inventory control methods, 'batch size of one' in production.

Very few managers take the opportunities open to them in this way—perhaps because they believe that surface mount technology is solely a new design and manufacturing issue and are unaware of its major impact on all the other departments within the company.

2.2 Subcontracting: the benefits

S1 Spreading financial risk

In many professional electronic products, the components and printed boards represent 60–80 per cent of the factory prime cost. By far the most important element in cost management lies in control of these; they form the main elements of working capital.

The subcontractor can be helping to finance the customer's company by purchasing and holding inventory on its behalf, typically prior to and during the board assembly sequence. This benefit can be an overriding factor in 'make or buy' decisions for small and 'start-up' OEM companies.

S2 Choice of technology

Surface mounting is not one technology, but several—depending on the degree of miniaturization sought.

Using a range of subcontractors enables a wider spread of technology to be deployed to suit individual product needs. In this context, the restriction of having to make use of an in-house facility because the money has been invested may sometimes prove to be a negative factor.

Balancing this is the comparative ease of optimizing each design to suit established in-house processes and equipment.

S3 Reduced technical risk

By employing specialist subcontractors who have greater expertise specific to their own assembly technologies, technical risk to the customer's company can be reduced. Sometimes product liability risks may be devolved by subcontracting (cf. Section 17.5).

S4 Choice of supplier

It is surprising how many technical managers still fail to grasp the advantage of selecting their subcontractor before starting their in-house (or subcontracted) board layout (cf. Section 19.1).

There can be greater scope for low cost when competitive tenders are sought that are appropriate to circuit structure, application category and quantity required. Often subcontractors specializing in high-volume manufacture will offer very low prices for prototypes just to get in at the start of a major project. From experience, they know that if they make the prototypes they stand a better than 70 per cent chance of getting a production order.

2.3 Making in-house: budgeting for the costs

M7   Management training
It is of the utmost importance to provide management training at all levels and in all departments before moving into surface mount technology.

Many case-histories reported by independent consultants have confirmed that a major problem exists in making managing directors of OEMs using conventional assemblies realize that they and their fellow directors are the people most in need of training for the change. This is irrespective of whether the involvement in surface mounting is via manufacture in-house or subcontracting.

Apart from the technology itself, there are significant differences in planning procedures and management control systems compared with those used in conventional assembly. Failure to appreciate these can severely prejudice the chances of success in applying surface mount technology.

The most important changes occur in the philosophy and control of design; in procurement, storage and kitting; in assembly processes, inspection and test. It is the sheer breadth of operational change right across the company that demands the attention of top management long before any plans for conversion are implemented.

M8 and M9   Capital investment and associated space and services
For a professional facility capable of manufacturing reliable products, the list and purchase price of capital items will almost certainly be greater than many people have been led to believe. Table 2.2 shows a typical item list, and if funds are limited Table 2.3 indicates a series of recommended distributions of the available finance.

For example, to cover all the company's circuit types on an economic 'horses-for-courses' basis, it may be necessary to buy equipment covering two or more of the basic soldering methods, e.g. flow/wave, vapour phase, infra-red, convection, thermode. To generate £1 million value of assembled boards, the capital budget for any two of these is likely to exceed £50 000. Soldering is the key process in surface mounting, and, for example, any company that spends less than £18 000 on a production infra-red soldering machine is likely to risk multilayer chip capacitor field failures within six months of product shipment, because of thermal shock problems.

Except for high-volume manufacture of single circuit types, production planning and return on investment for pick-and-place machines should be based on a prudent 70–75 per cent utilization figure and operation at 50 per

KEY FACTORS IN 'MAKE OR BUY' DECISIONS

Table 2.2 Key capital investment items, printed board surface mount assembly technology

Range: Millions of placements p.a.	£K 1–10	£K 10–100
1 Goods inwards inspection		
Passive components: Electrical test	1–5	10–25
Active devices: Electrical test		
Printed board: Bare board test		0–25
Counter for component reel/tube/tray contents or accurate weighing m/c		
Microscopes		0.5 each
Solderability test equipment	0.5	12–15
2 Cleaning equipment		
Components		0–10
Bare boards		
3 Printing		
Adhesives		0–15
Solder paste	7–10	10–50
Screen cleaning		5–7
Refrigerator	0.15	0.25
4(a) Component placement (pick-and-place automatic)		
Single head, serial pick-and-place, basic m/c	25–40	40–60
Twin head, serial pick-and-place, basic m/c	80–100	100–150
or Multi-head carousel, basic m/c		300–400
4(b) On-line feeders		
Tape reel feeders, 8–56 mm	0.3–0.6 each	
Stick/tube feeders	0.2–0.6 each	
5 Component placement (assisted manual placement)		
Simple pantograph	5–10	
Light spot indication	10–15	10–25
Programmable, restricted access	12–18	12–30
6 Adhesive dispensing and curing		
Single-head independent syringe m/c	10–15	15–30
Belt oven	7–10	10–12
7 Reflow soldering		
12 in mesh belt, 4–5 zones	18–20	
18 in mesh belt, 5 zones		25–40
Adjustable width edge conveyor		25–50
8 Wave soldering		
12 in wide board system	18–20	
18 in wide board system		25–50
9 Thermode soldering/rework unit for high pincount packages		
Basic machine (incl. preheat section)	40–50	
Thermodes to suit each component type	1–5 each	
10 Break-out equipment		
De-paneller	0.5	1–2

Table 2.2 *cont.*

Range: Millions of placements p.a.	£K 1–10	£K 10–100
11 Cleaning equipment		
Water: batch system	15–25	
Water: in-line system (incl. ultrasonics)		35–250
or		
Semi-aqueous: batch system	25–30	
Semi-aqueous: in-line system		35–250
Ionic cleanliness tester		25–30
Surface insulation resistance test equipment		10–15
12 Manual soldering and rework equipment		
Hot gas pencils	200–400 each	
Miniature soldering iron kits	50–150 each	
Heated tweezers	150–200 per pair	
Hot gas rework unit (multi-jet) incl. tooling		20–30
or		
Infra-red rework unit (shutter type) incl. tooling		25–35
13 Visual inspection		
Binocular microscopes	0.6 each	
Measuring microscope	2–5	3–7
Training microscope	2–3	4–6
Laser scanning system		300–500
14 Marking		
Circuit status	1	1
Circuit date coding		3–4
15 Electrical test		
General	20–30	20–50
In-circuit ATE	30–50	100–500
Functional test	50–100	
16 QA testing		
Temperature cycling unit	15–25	30–50
Adhesion force gauges	1	1
17 Services, etc.		
Anti-static flooring	4–8	30–50
Benches (anti-static)	0.5–1.0 each	
Clean compressed air (oil-free)	3–5	5–10
Cooling water system	2–3	3–5
Fume extraction system	5–10	10–30
Overalls	0.5	1–2
Packing and shipping equipment	0.5	1–5

cent of catalogue rated speed. These numbers do not include allowance for production yield effects, which obviously will vary according to particular circumstances.

Surface mount components and assembly techniques are still developing and are likely to continue to do so well into the next century. The rate of obsolescence of machinery will remain high, and, in preparing budgetary financial statements and business plans, companies should try to depreciate their assembly equipment over four years or less.

Services required for component placement, soldering and cleaning equipment are likely to include three-phase, 415-volt electrical supply with automatic peak-lopping, exhaust ducting, soft water, cooling water and specially filtered compressed air to prevent machine surfaces from contaminating components and fluxes. As a result of the rising tide of operator safety and environmental concerns, additional precautions and costs in disposing of toxic vapours and harmful waste liquids will be necessary.

A small production line capable of handling 1–2 million components per year may need 250–300 square metres of good-quality floor space, with a low dust count. The cleanliness requirement is not as stringent as for semiconductor assembly, but it is advisable to improve considerably on the standard of space housekeeping typically employed for conventional 'through-hole' assembly. A 'white-coat' environment is preferable, but to reduce the risks from static electricity, the coats should be made of cotton or, if nylon, should be specially treated each time they are laundered.

Post-soldering cleaning equipment and running costs are considerably higher and the advent of legislation banning use of CFCs will probably demand a manifold increase compared with the current levels needed for conventional component populated boards.

Table 2.3 Surface mount assembly: guideline capital cost distribution

Capital items	Available capital investment levels			
	£50 000	£100 000	£250 000	£500 000
Screen-printing	8	8	15	50
Pick-and-place: hand	7	9	15	25
Pick and place: auto	—	40	80	200
Soldering m/c: wave	18	18	25	30
IR	18	18	—	30
or convection	—	—	25	—
Cleaning: solvent	7	10	10	30
or water	7	10	40	50
Other items	10	15	40	100

Note: The above guidelines are for a typical small/medium-sized company with a wide range of circuit types. The distribution will be highly dependent on the application, but the advice is not to spend less than the recommended amount on the soldering machine.

M10 Technologist support

Because of the large number of variables in design, structure and, hence, processing requirements—especially if a variety of different circuits are envisaged—it is essential to budget for continuing technologist support on the production line. There have been numerous cases where management have assumed (unwisely) that the experienced technician in charge of through-hole wave soldering could understand the problems and handle the reflowing of surface mounted assemblies.

Qualified production engineers with materials and physics or chemistry background are required for this task. The average technician experienced in conventional soldering is unlikely to be able to cope without such support.

M11 Design rules and design training

This is the activity field where lack of training causes most money to be wasted by beginners. Successful design of surface mount electronic circuits and printed board layouts requires staff of high intellectual calibre as well as appropriate multi-discipline training and experience, plus, of course, good computer-aided design (CAD) equipment.

There is a tendency for those engineers and draftsmen experienced in layout for conventional boards to be over-confident when it comes to their ability to design cost-effective, reliable surface mount assemblies. Again, the greatly increased number of variables is the main hurdle, and mistakes can prove exceptionally costly. For example, when inevitably complex layout rules are not fully understood, or the space available for additional components or other changes is reduced as a result of miniaturization, complete rather than partial redesign becomes likely.

There is also an inherent reluctance to learn enough about printed board manufacture and the production assembly process to design the printing screen photomasters, and this problem then tends to be left to production to sort out. Invariably, the result is several successive screen redesigns as the production department struggles to get things right by trial and error.

In some instances, not only in military but also in consumer applications, elementary stress analysis is an essential part of the design procedure. It is important to have this skill available.

Sound design rules, validated by significant production experience, are still hard to come by for small and medium-sized companies just embarking on the new technology. Product cost and long-term reliability are highly dependent on these. Unless an experienced surface mount layout engineer is available within the company, purchasing a set of design rules from a specialist assembly operation may prove the least costly option. The data on pad layouts for individual components are likely to form only 10 per cent of the required information.

Independent CAD bureaux offering non-validated design rules and lacking short-term feedback loops from a 'live' production line are less likely to provide satisfactory layouts. This fact has been exemplified in a number of case-studies where, by tradition, the OEM has always asked a friendly local bureau to do their conventional 'through-hole' board designs. Following the same route for surface mounting has too often brought technical and financial mayhem to the project. The third case-history in Chapter 20 illustrates this situation.

If validated rules are offered, make sure they are not pirated. Some professional surface mount assembly houses entrust their rules to a bureau on the basis of signed confidentiality agreements and could take legal action if detectable breaches occur.

M12 CAD equipment
CAD equipments designed before 1992 are unlikely to prove complete and cost-effective for OEM total surface mount layout needs, despite the claims and 'bolt-on goodies' offered for earlier systems. Routing engines that are not yet sufficiently intelligent and the absence of simple thermal warning systems are among the absent features. The inevitable 'work-arounds' are fertile sources of error, but the situation is improving.

M13 Computer-aided design, manufacture and test (CADMAT)
Linking of printed board layout CAD outputs to assembly and test hardware is in its infancy at the moment. In the USA, the Surface Mount Equipment Manufacturers' Association is attempting to provide standard interfaces, but apart from linking to machines there is still a lack of simple software in the form of macro programs which could greatly assist the production engineer in both planning and cost estimating. Typical examples include the provision of pincount data to facilitate the costing of visual inspection and rework using synthetics, component count to give operating times on pick-and-place machines, and component style counts giving yield data for purchasing requirements.

M14 Component sourcing
Although the range of components available in surface mountable format is now very wide, the company's engineers and purchasing staff will need to spend above normal time seeking satisfactory sources for surface mounted components at competitive prices—at least for the next few years.

Currently, many of the surface mounted components being supplied to smaller companies are coming from distributors. The manner in which these lower quantities (e.g. <1000) are being stored, packed and shipped to users has not yet reached a satisfactory standard and assembly yields are adversely affected. Often this arises from poor solderability which has resulted from minimal solder coating thickness combined with unsuitable storage conditions and slow stock turnover.

M15 Production control
Both stock monitoring and circuit kitting for auto-onsertion machines—which can be fed from a variety of hoppers, chutes, trays, tubes and tape reels—present several interesting new complex control and accounting pitfalls compared with conventional assembly. These include the following:

- Many types of surface mounted component are very small and get dropped or lost easily when removed from their presentation packaging. Significant irregular losses can occur during machine set-up for each new circuit.
- Short lengths of tape from a reel may be needed for hand replacement of faulty components found during post-soldering visual inspection and test.
- The rough guides on the sides of the reels as to the number of components left on the tape are unlikely to be accurate enough when individual item costs are high; e.g. for resistors priced at less than 0.5 pence each this is unimportant, but for capacitors or semiconductor devices at £1+ each, it may be a different matter.
- Many chip components still do not carry indentity marking as to their value, tolerance, date code or source. This means that loose items should always be stored in a manner that enables their key parameters to be determined easily.
- As a result of the small size of chip components and the reflow assembly techniques employed in mounting them, the solderability of their terminations and the respective printed board footprints is of much greater concern. It becomes vital to monitor shelf life and impose a rigid discipline on 'first-in/first-out' (FIFO) stores procedures. Preferably these should be based on 'date of manufacture' rather than 'date of receipt', i.e. adopting the 'earliest date-code in/date-code out' (EDI/EDO) principle.
- Revised methods of control are needed to monitor the movement of part-empty reels, tubes and trays after a batch run is completed. It is important to prevent them from being shunted off to a private store for use in a 'rainy day' poor-yield situation!

M16 Operator training
Operators familiar with through-hole assembly will need considerable training times to accommodate to surface mount standards. Apart from the use of more sophisticated machinery, the need is for a high level of manual dex-

terity and good eyesight. Colour-blind people should be offered alternative work. The manipulative skills are required mainly in connection with prototypes and the assembly of 'odd-ball' components, but also for rework on products assembled automatically. Training periods ranging between 100 and 1000 hours are likely, depending on the required quality and complexity of the work.

An important activity needing periodic retraining is that of visual inspection, where the wide variety of surface mount component package styles and terminations, the reduced clearances on the printed board and the much smaller solder joints combine to make the task more intricate and tiring.

Sufficient intellect must be available to cope with unmarked chip components, particularly on short-run jobs. Diagnostic probing by hand to search for electrical and mechanical faults often needs a combination of high electronic intelligence plus expert finger skills. This can be expensive and puts a premium on 'in-circuit' testing where each individual component is exercised, giving instant diagnosis and effective location of defects.

M17 Start-up losses
These losses stem from many causes. From previous experience they have included:

- Extra CAD, photographic and printed board tooling costs arising from learning curve design errors
- Incorrect printed board specification leading to unsuitable materials and manufacturing processes being used by suppliers
- Major production yield losses during the shop-floor learning curve period
- Additional costs in assembly from incorrect and delayed component supply
- Inadequate testing due to insufficient operating test nodes being designed in
- Unwise decisions by inexperienced inspectors, often resulting in excessive rework and generating reliability problems
- Poor machine utilization
- Diversion of scarce management resources

M18 Staying in the game
The technology and its associated materials and components are all developing rapidly. Be aware that an initial capital outlay may need to be followed within a year or two by a second tranche—essential if the company is to stay in the game on a competitive basis.

22 SUBCONTRACTING ELECTRONICS

M19 Exit phase
This heading may seem inappropriate, but experience has shown that the likely cost of closing down either a successful or an unsuccessful OEM investment in an in-house technology project should be considered at the outset.

2.4 Making in-house: a disadvantage

M20 The 'technology stretch' syndrome
All too often, engineers working for in-house assembly operations tend to get pushed near or over the edge of sensible economic technology limits by enthusiastic system project leaders who are keen to squeeze the utmost from circuit performance and/or size reduction.

Almost invariably, this results in unforecast high production yield losses and programme delays due to manufacturing difficulty. Often major problems arise when attempts are made to source externally at a later date. Professional subcontractors are usually unwilling to follow marginal design and manufacturing practices.

As mentioned at the start of this chapter, an important feature of today's electronic scene is the steadily shortening life-cycle of products in the market-place. This can seriously erode a company's ability to recover major investment costs in new products, and if we then add the finance and resources needed to set up new technology in parallel, the option of subcontracting could appear sensible.

Conversely, it is argued that shorter product life-cycles can be accommodated only by setting up an in-house facility in advance so that at least the time spent in seeking reliable sources and negotiating with subcontractors is eliminated.

For large OEMs, the relevance of installing their own surface mount technology will be obvious to them. However, in small and medium-sized companies the capital sums, though comparatively large, are likely to represent less than half the overall start-up costs, and their 'make or buy' decision should be prefaced by an in-depth cost–benefit analysis.

2.5 Subcontracting assembly: budgeting the costs
This section outlines some of both the obvious and less obvious costs inherent in subcontracting surface mount assemblies.

S5 Full specification
To ensure an efficient commercial and technical interface with a subcontractor, the company will need to put in writing far more than would be necessary for in-house manufacture and certainly more than for subcon-

tracted conventional assembly. Time lost in obtaining quotes and in ensuing technical discussions cannot be recovered.

S6 Second sourcing
If second sourcing is needed, as well as repeat tooling and additional text hardware and software charges, the cost of engineering time in discussion and data transfer will be proportionally higher.

S7 Production runs
Unless otherwise encouraged, for small companies subcontractors may not find it economic to set up an auto-placement machine programme for the prototypes: they will use hand assembly methods. Hence the decision to subcontract or make in-house should not necessarily be linked to early models of the product. A subcontracted production run of 1000 units or more may be needed to enable valid long-term 'make or buy' decisions.

S8 Free-issuing components
If it is intended to retain close control of material supplies and to send all items to the subcontractor on a 'free-issue' basis, the additional costs in kitting, packing, shipping and arguing the toss about shortages and quality must be carried by the company. These may easily offset marginal purchase price advantages.

S9 Loss of added value
The subcontractor company is taking added value, but also is adding to its investment risk. It may therefore feel entitled to a higher margin than hitherto to obtain an adequate return on its investment in a new, capital-intensive technology.

2.6 Subcontracting: some disadvantages

S10 Late delivery risk
Given a sticky output situation, some subcontractors may tend to put their production priorities where good gross margin or high added value are most easily attained. Murphy's Law applies: your product is certain to be among the unlucky ones!

S11 Testing
The subcontractor's test capability and knowledge of a customer's specific circuits is unlikely to match that of the OEM's own staff and equipment. This

is critical where fault-finding is needed for complex boards. In-circuit testing by the subcontractor can cover most of the problems, but it cannot replace final functional test. Ideally, the subcontractor should do both—provided the extra hardware and software costs can be afforded by the OEM.

Visual inspection has been aptly described as being '80 per cent efficient for 10 minutes'. Calling for visual inspection alone is unsatisfactory as a basis for an OEM to receive complex circuits from a subcontractor. At 200 components or 700 solder joints per board and a defect rate of 500 parts per million (p.p.m.), it can easily be shown that every single circuit will have at least one fault on arrival. The ensuing shuttle service between supplier and customer is of benefit only to the carrier companies.

A logical solution in this case is for the subcontractor to help customers to set up their own small surface mount rework and repair stations and to negotiate a suitable acceptance arrangement and a price adjustment formula. In this way any field returns can also be dealt with by the OEM instead of having to be passed on.

2.7 Conclusions

Competent managers in different departments of a company can usually make out a 'cast-iron' case for either of the 'make or buy' options considered above, depending on their interests or ambitions.

The difficulties in this situation are often overcome by employing an independent consultant. The problem is to find one who combines sufficient first-hand knowledge of surface mounting with a good appreciation of the company's business and products.

If the decision goes in favour of subcontracting, ensure that the company's engineers and purchasing staff understand the new style of relationship needed (cf. Chapters 13 and 17). For many reasons the time-scale between order placement and receipt of product tends to be longer than for conventional assembly, and the interface with the contractor needs to be closer and more detailed—also more friendly.

If it is decided to purchase and install a surface mount production facility within the company, ideally the project leader should be the MD, or at least someone at senior director level. The selected person must have received appropriate management and technical training for the co-ordination tasks. The reason for this suggestion is evident from M5 above. The new technology will impact on the operating methods of every department in a manufacturing company, and the appointment of a less senior person has been known to be a recipe for internal chaos.

A logical compromise for a small company is to set up a small in-house prototyping capability to enable rapid sample manufacture, but to plan to subcontract all production right from the start. If things go well, the prototype line can be expanded at a later date.

This solution carries with it several important provisos:

- There is a need to select a suitable subcontractor in advance so that appropriate printed board layout rules, machine requirements and process sequences can be agreed before any circuit design and board layout work commences.
- For surface mount assemblies it is vital that type approval testing reflects the processes used in production. Carrying out this work on prototypes made under laboratory conditions may give significantly misleading results—either way!

An extension of the previous option is to plan limited in-house production and buy outside for the overflow requirements. Many experienced subcontractors are wary of this situation and may not be too keen to accommodate such activity unless it is on a long-term, stable basis, e.g. year by year.

3

Contracts, Terms and Conditions of Purchase and Sale

3.1 Types of contract for board assembly

There are three basic levels of activity to be covered by contracts between customers and their board assembly subcontractors. These are: prototype, pilot production, and continuous production phases.

In cases where there is no formal contract drawn up between the parties, the legal position will be based on six documents, which may not necessarily be in total agreement. These are:

- The customer's confidentiality agreement document
- The subcontractor's quotation
- The subcontractor's Conditions of Sale
- The customer's Conditions of Purchase
- The customer's purchase order and attachments
- The subcontractor's acceptance of the purchase order

Only in major, long-running projects is there likely to be a separate contract drawn up and signed by both parties.

PROTOTYPES

While customers will want to have straightforward firm prices and delivery statements, professional subcontractors are likely to seek protective clauses which impose conditions on what is to be delivered and qualifying phrases closely defining and limiting their responsibilities.

Often customers will request a target production price at the same time as they are asking for a price covering prototypes. Here again, there will be a qualified response from the experienced subcontractor and the customer will have reason to be suspicious if this is not so.

Most subcontractors are happy to quote a firm price for prototypes, but will prefer to give only budgetary ones for subsequent pilot and full-scale production quantities—and even then, they are likely to be hedged about with 'ifs' and 'buts'. Experience has shown that a customer expecting more than this is requiring an act of folly from the subcontractor.

An idea of what one should expect to see in a professional quotation for prototypes and a basis for judgement is given in Section 13.3.

PILOT PRODUCTION

Subcontractor statistics suggest that, if a prototype order has been received and fulfilled, then there is a 60–70 per cent chance that the same supplier will receive at least a healthy proportion of the first production order.

It is to be hoped that the product, when given its first pilot production run, will have had the benefit of design improvements; nevertheless, professional subcontractors will still wish to insert protective clauses which impose conditions on what is to be delivered and qualifying phrases, again closely defining and limiting their responsibilities.

FULL PRODUCTION

Apart from price and quality, two of the most important characteristics of a full-scale production order are the ramp-up phase and the permitted fluctuation levels in ongoing shipments.

Experience also shows that excessive OEM pressure for rapid completion of the project, combined with subcontractor optimism (which, translated, means 'keenness to get the order'), a willingness to believe component supplier's delivery promises and over-confidence in an ability to move mountains, usually result in mutual disappointment. Where practicable, it is better if the customer asks the subcontractor company to set its own ramp-up rates and dates and then (if possible) enforces a slightly reduced programme based upon them. Those who believe this concept to be approaching heresy may wish to consider carefully the meaning of the word 'partnership'.

The same should apply to the allowable fluctuation percentages and the associated minimum advance warning times. Where subcontractors see a risk that call-off rates are likely to be significantly lower than the customer says (they nearly always are), they will want to build in a price adjustment clause to take account of reduced quantity throughput and any consequent inventory holding costs. Refer to the sample production quote given in Appendix 13B.

Customers will aim to have such clauses framed as an adjustment to the annual or final project settlement rather than relating price variations to current shipments. Subcontractors will be negotiating for short-term adjustments to assist their cash flow situation.

3.2 Terms and conditions of sale and purchase

When there are divided responsibilities for the quality and reliability of a product, the conditions of sale and purchase obtaining between the subcontractor and customer must reflect them in a well-defined manner.

If things go wrong and the respective responsibilities are not clearly laid down, are not determinate, or are collective, a fair partitioning of liability is needed. One option applied by arbitrators is to apportion them on the respective gross margins of the parties in relation to the product at issue.

It is unlikely that the normal conditions of purchase used by an OEM to procure components or boards will be satisfactory, and if there is no action to amend or augment them to suit the circumstances, those of the subcontractor may apply by default.

At the same time, the conditions applied by some OEMs are sometimes unreasonable in demanding, for example, unlimited consequential liability, or in stipulating that no liability can arise from their own delays in delivering parts or postponing delivery schedules in a way that imposes rescheduling costs and/or extra cash flow problems on their suppliers.

One solution is to negotiate a formal compromise contract covering the planned business between the parties. This can combine their key concerns, but is usually based on the purchaser's Conditions of Purchase. Some typical contract terms are given in Appendix 3A. They are self-explanatory and are not intended to be exhaustive.

In addition, a typical set of Terms and Conditions of Sale from a professional subcontractor is given in Appendix 3B. These also are not necessarily exhaustive, and some of the salient points to watch are as follows:

- Subcontractors will want to be free to pass on any unforeseen increases in the cost of materials or labour that occur between their quotation dates and shipment of the goods, including any increase that may be caused by the action of their customers. This clause will be worded to allow for customer-induced delays as well as extra costs arising from supplier sources, design or process changes and increased difficulty in manufacture.
- Some flexibility in the delivered batch quantities should be built into the delivery schedule. This will be intended to take account of the variable output yields which can occur in mechanized surface mount assembly. A usual range allowed would be ± 10 per cent. For long-term projects there are technical arguments in favour of this principle which can benefit both

parties. As far as the customer is concerned, if there is a period of more than three months scheduled between successive deliveries, the solderability of contact areas on the assembled board will be reduced and it may pay to accept and use the entire manufactured batch immediately rather than leave a few excess items in the subcontractor's stores for later delivery.
- If the subcontractor is to be responsible for procurement of part or all of the product materials, some payment in advance may well be expected. If this is not practicable, an increase in price to cover the extra risks may be expected.
- As far as the subcontractor's design and tooling costs are concerned, it is usual for these to be invoiced at time of shipment of the first batch of product made using them. When subcontractors describe these costs as 'part-tooling charges', this means that they are laying claim to their ownership and are trying to reserve the right to keep them after the contract is completed. If they have made the layout, then it will undoubtedly contain an element of their intellectual property, and, should the customer wish to have the right to retain the photomasters or disks and tapes containing data thereon in order to use them with a second source, there could be problems. These should be ironed out within the initial contract. A typical solution is for the subcontractor to agree to hand over all data in return for an additional sum of money.
- Subcontractors will want to be 'held harmless' and absolved from liability for defective product performance arising from the customer's own designs and procurement specifications, including any health and safety issues. They will also seek to avoid all consequential liabilities (see clauses 8(c) and 6(h) in Appendix 3B). In some instances, for example in medical applications or usage in the USA, the subcontractor will want to be held harmless regardless of blame. In countries where 'strict liability' applies, this is not necessarily a protection against attack by a harmed party.
- When the subcontractor has reason to suspect that electrical testing by the customer or test equipment specified or supplied by the customer is causing defects ascribed to the subcontractor, the subcontractor may want to include right of access to the customer's test equipment as a condition of sale.
- The subcontractor will seek to make the customer responsible for all free-issue items at all times, including during storage and use on the subcontractor's premises. Conversely, the customer will wish to make the subcontractor responsible by claiming that this is normal trade practice. The question of who should insure the free-issue items must be settled within the initial contract terms. Either way, the customer must expect to pay.

3.3 'Turnkey' projects

In the context of subcontracting in the electronics industry, 'turnkey' projects are those in which the subcontractors handle the completed product rather than just board assembly. In so doing, they also perform many of the normal functions of their OEM customers, e.g. purchasing, stores, manufacturing, quality, test, packaging and, in some cases, storage and dispatch direct to the customer's customer. Typically this would leave the OEM with design, marketing and sales functions.

The simple truth is that in many turnkey arrangements subcontractors are being asked to make a major investment in the project. They are therefore entitled to at least the same financial protection and security status as any bank that lends money to it—a situation that is seldom realized in practice. Both parties are expected to explore the financial probity of their potential partner in greater depth than for a mere board assembly contract.

THE RISKS

A key financial issue in this type of arrangement is the fair distribution of margin according to added value and risk. Those subcontractors who approach a turnkey project with the same ground rules they adopt for board assembly business could suffer badly. Apart from the consequences of design changes, the main extra risk for them lies in the size of their inventory holding and consequent cash flow exposures—especially if they are asked to maintain sufficient stocks of completed product in their finished-goods stores to supply on demand and direct to the end-customer.

The main risks for the customers are that neither the output nor the quality of the product are under their direct control and, should they get into financial difficulty, the subcontractors can simply hang on to stock unpaid for.

For the subcontractors, their profit will be far more dependent on the success or otherwise of their customers' marketing and sales departments. Clearly, the greater interdependence of both parties means that at least the final negotiations for the deal should be at board level, even when the customer is a large OEM.

Before reaching this stage, obviously the customer will carry out a very thorough audit of the subcontractor's company and its *modus operandi*. To this end, previous experience of its performance on board assembly and testing are valuable indicators (cf. Chapters 14 and 16).

Similarly, before accepting a major project, the professional subcontractor will want to take a good look at the customer's sales and marketing functions, e.g. to explore their organization, calibre, planning and coverage. OEMs should be prepared for and accept this request as an essential part of the deal and one that will be beneficial to the project (cf. Chapter 14).

STOCK AND VALUATION

In case of discussion or dispute, it is prudent to include agreement on the following subjects prior to signing any contract.

Pre-process material stocks held by the subcontractor

Normally the cost of holding pre-process stock would be included either in the overhead or in the subcontractor's material handling charge. To ensure that subcontractors obtain sufficient added value and to cover their expenses in procurement, inspection, storage, kitting and logistics, either way they will need between 8 and 12 per cent of the total materials cost above the current bank lending rate.

Because the range of material costs as a proportion of total costs can vary widely between projects, it is fairer to the individual customer to avoid putting them into the overhead.

Work-in-progress (WIP) valuation

In discussing the value of obsolete or unusable WIP, it is important to record in the contract that the customer accepts the subcontractor's normal method of monthly valuation or, if this is not possible, to agree in advance an alternative for use in the event of a change in sourcing policy or other difficulty.

A simple method adopted by some subcontractors is to assume that the value of the total WIP is half the total prime cost of all the circuits in production at the end of each accounting period. This, of course, is a serious underestimate for board assembly work because the bulk of the cost is in the components. However, it will be slightly less inaccurate for complete products. On the same simple basis, a fairer way is to take, for each product, the yielded materials cost (including the handling charge) plus half the estimated added value.

For the larger, more sophisticated subcontractor, one method is to partition the growth in cost as the product moves through each successive process; often the figures are expressed as percentages of its final prime cost. For example, starting with the percentage value of materials, one could see a progression typically as follows:

Ready kits of parts	60%
At printing solder paste through to after component placement	70%
After soldering, awaiting rework	75%
After rework, awaiting or at visual inspection	85%
After visual inspection, awaiting electrical test	92%
After passing electrical test	98%
After packing, awaiting shipment or in finished-goods store	100%

This type of system is practicable only when the on-line data collection of product status has reached a highly professional standard—usually involving remote input stations on the production line connected to a central production control computer. It also requires the cost-estimating software to include in the standard cost-estimate sheet printout the relative percentages at each of the key process stages (cf. Chapter 19).

Finished goods stock
Valuation presents no mathematical problems at this stage. Subcontractors will expect to receive payment in the normal way for all product transferred either to the customer or into the agreed finished-goods (preferably bonded) store on their own premises.

However, subcontractors will wish to reach agreement with their customers on the price for holding stock on their behalf—in percentage terms related to the product cost (or selling price)—and also on the mechanism for calculating the invoiced charge for this service on a monthly basis.

One of the reasons why professional subcontractors would want to deal with this subject on a separate basis is if they are operating the usual good practice of treating all finished custom-built goods as having zero value. If they do have a few overmakes through better-than-forecast production yields, there is no guarantee that further orders will follow and the practice is therefore only prudent. In any event, the level of finished goods is almost always negligible and will not have been built into the subcontractor's normal business overhead calculation.

By far the most critical aspect that must be thought through by the customer is what happens if either party hits financial trouble. If the customer has this misfortune, the subcontractors are in a relatively strong position if they hold the stock, compared with the situation where it is held by the customer and has not been paid for. This is one reason underlying clauses 7(a) and 7(b) in the subcontractor's Terms and Conditions of Sale (see Appendix 3B).

If it is the subcontractor who fails, the customer will be able to enforce collection only of free issues and finished goods already paid for. All other stock will require negotiation, and it is in this situation that prior agreement on the method of valuing the pre-process materials and WIP can prove invaluable.

PRICE AND QUANTITY VARIATIONS
Subcontractors will consider it more important than usual to negotiate retrospective price adjustments reflecting manufacturing output quantities. This can easily be managed on a rolling basis, but may mean that they will

need to demand a similar approach from their component and board suppliers.

COST-REDUCTION PROJECTS

For long-running projects, early discussion on the principles to be adopted in the continual pursuit of cost savings and the reflection of these in suitable contract clauses is desirable.

It is good business practice to encourage subcontractors to suggest cost reductions on an ongoing basis, notwithstanding their potential loss of profit. Where these involve their own investment in changes for handling and process methods (cf. Section 17.3), it is reasonable that they should have the option of retaining any benefit or passing some or all of it on to their customers.

When cost-reduction projects involve design or other changes requiring the customer's approval and some mutual expenditure, the case for sharing the benefit to offset the subcontractor's loss of profit is clear. The method of reward should have been agreed in advance.

3.4 Arbitration

In this field, independent auditing or arbitration are often better ways than normal court procedure to resolve disputes between parties. While the cost of professional services is unlikely to be less, there can be benefits from increased background knowledge. The decisions by an independent expert are likely to be based on a better understanding of the technical issues than those of a judge in court.

Clearly, the choice of expert is critical. He or she must have knowledge of key legal and liability issues as well as the appropriate specialist technical expertise. Often it is a matter of finding the right source of help. For example, in the UK the Surface Mount Club or the Electrical Research Association might be contacted for advice. A list of possible sources is given on page 259.

Appendix 3A Example Agreement Terms between an OEM and a Subcontractor

1 Definitions
"Conditions" shall mean the conditions of this Agreement and of any Contract placed thereunder for purchase of Goods by SSC from GSL;
"SSC" shall mean Sensible Systems Corporation;
"GSL" shall mean Guardian Subcontractors Limited;
"Contract" shall mean the specific terms agreed in relation to an Order placed under the general terms of this agreement;
"Order" shall mean SSC's written instructions to supply goods and services in a Contract placed under the general terms of this Agreement;
"Goods" shall mean the goods and services embodied in the Order;
"Parties" shall mean the parties to this Agreement and to any Contract thereunder;
"Quotation" shall mean any offer by GSL to supply Goods and Services under the terms of this Agreement and any Contract thereunder.

2 Order Acceptance
(a) All Orders placed by SSC under this Agreement will be acknowledged in writing by GSL within 5 (five) working days of receipt of the Order or amendment thereto.
(b) Verbal instructions given by SSC to GSL shall not affect the contractual obligations of GSL unless they are confirmed in writing to GSL within 5 (five) working days of their being given.
(c) Any Order acceptance from GSL shall be deemed to constitute an acceptance of and agreement to comply with the Conditions.

3 Conflict
(a) All express conditions which conflict with the Conditions are hereby excluded.
(b) In the event of conflict between the general terms of this Agreement and any Contract placed thereunder, the terms of this Agreement shall prevail.

4 Design and Quality
(a) Responsibility for all design matters shall rest with SSC, including the specification of all items for incorporation in the Goods and the fitness for purpose of the design over the planned life of the product(s).
(b) GSL are responsible for the quality of the assemblies, that is to say for the quality of all non-free issue parts and materials used in the prod-

ucts, for the correct components being in their specified positions and orientations, for the quality of all solder joints made by GSL, for the processes and cleanliness of the assembly, for the conformance of the products to agreed mechanical specifications and outline dimensions and for the carrying out of such electrical or other tests as are specified in the Contract.

5 Control of Changes
(a) No changes shall be made to the design, whether they be electrical, mechanical, materials or other, without the agreement in writing of SSC.
(b) Should either party propose a change in the product for whatever reason, details of the proposed modification shall be made to the other party prior to its implementation and consent shall not be unreasonably withheld. If appropriate, an equitable negotiated adjustment shall be made to the price of the affected products.

6 Inspection of Product during Manufacture
(a) GSL shall not unreasonably withhold permission for SSC staff and bona fide SSC customers for the specific goods made by GSL to have access to the relevant GSL and assembly and test facilities at mutually convenient times for the purpose of monitoring the tests prescribed in the specification.
(b) GSL shall, if requested, supply copies of all test reports and inspection documents as may be reasonably required for verification of the process control and quality of the products.

7 Rejection of Goods by SSC
(a) When Goods are rejected by SSC, SSC shall return them to GSL and within 15 (fifteen) working days provide detailed information as to the location of faults and the reasons for rejection.
(b) At the request of SSC, GSL shall, at their risk and expense, quickly repair or replace all Goods mutually accepted as reject.
(c) All repaired or replaced items shall receive specified testing as appropriate to their condition and status on the assembly line and the Conditions shall apply to all such Goods.
(d) SSC shall have the right to cancel the Contract in whole or part if, for any returned product accepted by GSL as defective, GSL fails to provide repaired or replacement product within 15 (fifteen) working days of receiving the returned items or the defect information thereon, whichever is the later.
(e) Clause 7(d) above will not apply if the delay arises from Force Majeure

or from failure by SSC to effect timely supply of any necessary free-issue items.

8 Disputes

(a) All disputes or disagreements of whatever kind notified by either party shall be discussed between the parties and settled amicably.

(b) In the event of a dispute as to whether the product conforms to the specification or which party is responsible for causing a defect or defects or for late delivery or any other disagreement, either party may refer the question to the Surface Mount Club at the National Physical Laboratory, Queens Road, Teddington, Middlesex, TW11 0LW, UK, whose decision shall be final and binding.

(c) Should either party notify its intention to appoint an alternative auditor and failing agreement of the parties on the proposed alternative, the dispute shall forthwith be referred to an arbitrator nominated by the London Court of Arbiters and any such submission shall be deemed a submission within the meaning of the Arbitration Act 1950 or any amendment thereof for the time being in force. The decisions of the arbitrator so appointed shall be binding on both parties.

(d) During the period of investigation by the appointed independent auditor or by the arbitrator, the Contract shall continue as if no dispute existed.

(e) The cost of such referral shall be paid by the parties in such proportions as are decided by the appointed auditor or arbitrator.

9 Title and Risk

(a) Excepting those items rejected under the terms of Clause 7 or as agreed in writing, ownership and risk in the Goods shall pass to SSC on delivery.

(b) Ownership and risk shall return to GSL within 5 (five) working days after Goods claimed as reject by SSC, or the defect data thereon, whichever is the later, are received by GSL.

(c) All designs, patterns, drawings, software, tools, test jigs and test equipment and production tooling paid for or provided by SSC shall be and remain the property of SSC.

(d) Items part-funded by SSC shall, at their request, be sold to SSC for an amount not exceeding their book value less the depreciated value of the SSC contribution.

10 Process Information

For the period in which GSL remains as sole supplier of a product or at closure of the Contract GSL shall, on request, supply to SSC all process documentation applied to the product.

11 Guarantee

Without prejudice to the rights of SSC under the Contract, GSL shall guarantee the Goods for 18 (eighteen) months against defects arising from failure to fulfil their obligations and responsibilities under Clause 4(b) of this Agreement. Any defective Goods returned to GSL within the guarantee period and mutually agreed as defective shall be quickly repaired or replaced and prior to shipment be made to comply with the specification extant at the time of their original dispatch from GSL.

12 Traceability and Identification

GSL shall mark and pack the Goods in accordance with the SSC specification and shall keep records enabling:
(a) Traceability of all items in the product as to their country of origin, manufacturer and delivery date to GSL. (In the case of free-issue items supplied by SSC, only the delivery date shall be recorded.)
(b) Confirmation that they comply with their specifications.
(c) The serial number of each product shipped to be traceable to a Delivery Note number.

13 Delivery

Time is and shall remain of the essence of this Contract.
(a) The Goods shall be delivered as to date and place as specified in the Contract. Deliveries earlier than specified in the Contract require agreement in writing from SSC.
(b) On or before the first day of each calendar month, SSC shall provide GSL with a forward rolling delivery schedule covering weekly deliveries over the ensuing four months. Deliveries scheduled over the first three months shall be fixed and shall constitute the current contractual requirement for both parties. Deliveries in the fourth month may be subject to alteration by SSC. Such changes shall take into account the forward materials order commitment of GSL on behalf of SSC and any cost penalties imposed by suppliers to GSL shall be for the account of SSC. Nevertheless, GSL shall use its best endeavours to minimise the potential of resulting exposures and penalties when placing related forward orders and when notifying its suppliers of such changes.
(c) When such changes in the schedule or the non-delivery of free-issue parts from SSC result in production stoppages causing additional

machine or labour costs to GSL, GSL will use its best endeavours to provide alternative work and in any event shall not charge more than £150 per hour for such periods of stoppage. Where resulting charges are claimed by GSL they shall be invoiced not later than 60 (sixty) days after the events charged for and SSC shall have right of access to relevant operator and machine work records. This access shall not constitute reason for delay in payment under Clause 14 of this Agreement.
(d) If GSL fails to deliver in accordance with the schedule agreed in the Contract, subject to the conditions given in Clauses 7(d) and 8(c) above, SSC shall have the right to cancel the Contract in whole or in part.

14 Force Majeure

(a) When either party is unable to meet the terms of this Agreement or a Contract thereunder by reason of circumstances beyond its reasonable control, including but not limited to strikes, lock-outs, labour disputes, armed conflict, civil disturbances or riot, fires, floods, acts of God, acts of Government, currency restrictions, unavailability or shortage of materials, breakdown of machinery or failure of a supplier, carrier or third party subcontractor to deliver on time, they shall be granted a reasonable extension of the time allowed to meet the required terms.
(b) The party affected by Force Majeure shall take all reasonable steps to minimise its effects and immediately notify the other party of any inability to fulfil the terms of this Agreement or any Contract made thereunder and advise on the reason(s) for its inability and the probable length of any resulting time delay.
(c) Both parties shall have the right to terminate the Agreement or any Contract made thereunder if the continuing effects of the Force Majeure, for example prolonged strikes, seriously impair the affected party's medium- or long-term ability to fulfil its commitments to the other party.

15 Payment

(a) GSL shall provide SSC with a monthly invoice covering all shipments, services and credits provided in the preceding calendar month. The invoice shall be accompanied by a list of all product serial numbers shipped and credited within the period.
(b) SSC shall effect payment within 30 (thirty) days of the end of the month in which SSC receives the invoice. Payment shall not constitute a waiver of any of SSC's rights under the Contract.
(c) In the event of non-payment, GSL shall provide notification to SSC of

its intention to withhold deliveries and may effect suspension of shipments after 5 (five) working days have elapsed since notification without being in breach of contract.

16 Communications
For the purpose of this Agreement and all Contracts operating thereunder, communication by letter, by telex or by facsimile shall be deemed to constitute notification in writing.

17 Indemnity
In relation to the respective responsibilities of the parties given in Clause 4 of this Agreement, SSC and GSL shall indemnify each other against any respective action, liability or expense caused by infringement or alleged infringement of any copyright, trade mark, registered design or patent relating to the Goods supplied to SSC.

18 Subcontracting
(a) GSL shall not subcontract any part of the Contract for assembly and test of the Goods without the written agreement of SSC. Such agreement shall not be unreasonably withheld provided the requirements of Clause 6 of this Agreement and the specification of the Goods are met and that the subcontractor is bound by the applicable terms and conditions of the Contract.
(b) Save under conditions of Force Majeure, subcontracting by GSL shall not relieve it of its responsibilities under this Agreement or any Contract thereunder.

19 Termination of Agreement or Contract
(a) Termination of a Contract under the terms of this Agreement shall not necessarily terminate this Agreement unless either party decides to do so.
(b) Either party may terminate this Agreement or any Contract thereunder by giving a minimum of 20 (twenty) weeks' notice in writing. Should SSC terminate the Agreement or any Contract thereunder, SSC will accept delivery of all finished goods, all pre-process materials and all work in progress at the date of termination and will pay a fair and reasonable sum for these goods. SSC will also pay for such items as are covered by Clauses 9(c) and 9(d). Upon receipt of such payments, GSL shall deliver the agreed items to SSC within 15 (fifteen) working days.

(c) Either party may terminate a Contract or part thereof or this Agreement if, after notifying the other party of the existence of a breach thereof, the said breach is not rectified within 20 (twenty) working days from the date of such notification.

(d) If either party becomes bankrupt or insolvent or has a receiver appointed of its undertaking or assets or a substantial part thereof or has any execution levied on its goods or assets or (being a limited company) goes into liquidation other than a voluntary liquidation for the purpose of reconstruction without insolvency, the other party shall have the right to cancel the Contract(s) made under this Agreement. Such cancellation shall be notified in writing to the other party or the liquidator or to the receiver or to anyone in whom the Agreement or Contract has become vested. Such notification shall not prejudice the existing rights of either party.

(e) Under the conditions defined in Clause 15(c) above, the other party referred to in Clause 19(c) above shall have the right to give the liquidator or receiver or other such person the option of continuing to operate the Contract up to a financial limit to be agreed.

20 *Enforcement*

Should either party fail to enforce any right conferred to it under this Agreement or conferred to it in any Contract made under this Agreement, this shall not be deemed to be a waiver of any such right at any time or times thereafter.

21 *Statutory Regulations*

Within the respective responsibilities defined in Clause 4 of this Agreement, SSC and GSL agree to comply with all relevant regulations, statutes, rules and by-laws and EEC directives affecting the performance of this Agreement and any Contract made thereunder.

22 *Confidentiality*

(a) All information including but not limited to processes, specifications, drawings, software, modifications, and other documentation used for the purposes of defining, assembling and testing products for SSC shall be treated by both parties in strictest confidence and its dissemination within respective companies shall be restricted to those who require such information for the proper performance of their duties.

(b) Neither party shall make any disclosure to any third party relating to this Agreement without the written agreement of the other party except as may be necessary by reason of statutory, legal, accounting or regulatory requirements.

(c) Both parties will take all necessary steps to ensure that its employees are aware of the need to maintain strict confidentiality, including the avoidance of risk that other visitors to the parties' premises shall, by observing documentation or labelling in production or other areas, become aware of the existence and nature of the business between the parties.
(d) These obligations shall survive the termination of the Agreement by 2 (two) years.

23 *Law*
This Agreement and any Contracts made thereunder shall be governed by English Law and, save any settlements reached under Clauses 8(a) or 8(b) of this Agreement, the parties hereby submit to the non-exclusive jurisdiction of the English Courts.

24 *Previous Agreements*
This Agreement supersedes all previous agreements between the parties.

Appendix 3B   Typical Subcontractor's Terms and Conditions of Sale

1 Definitions
"The Company" shall mean . . . (e.g. Subcon Limited).
"The Customer" shall mean the purchaser of the Goods.
"The Goods" shall mean all goods and services embodied in the Order.
"The Order" shall mean the Customer's offer to purchase the Goods on the Conditions.
"The Conditions" shall mean the terms and conditions contained herein.
"The Contract" shall mean the Company's written acceptance of the Order.
"Part tooling charge" shall mean a contribution towards the cost of tooling.
"Part design charge" shall mean a contribution towards the cost of design.
"Delivery" shall be ex-works.

2 General
2a In the event of inconsistency between the Conditions and those of the Customer or any other party, the Conditions will prevail.
2b Any variation in the Conditions shall be made in writing and signed by an authorised officer of both parties.
2c Any forbearance shown by either party in respect of any of the Conditions shall not be a waiver of any rights under the Conditions.
2d The Customer shall not assign the Contract or any rights thereunder without the written consent of the Company.
2e The headings in the Conditions are for convenience only and shall not affect the construction thereof.

3 Quotations and Prices
3a Unless otherwise specified by the Company in writing, quotations are subject to materials being available on receipt of Order and subject to Clause 3b of the Conditions valid for 30 (thirty) days from the date of quotation.
3b The Company reserves the right:
   (i) to pass on to the Customer increases in costs to it after the date of the quotation and before delivery including (but not limited to) increases relating to materials, wages, currency exchange rates, transport and taxes, or where the increase is due to any act or default of the Customer including (but not limited to) the supply by the Customer of delayed or incorrect or faulty components, printed boards or piececparts or of test equipment which is slower to operate than anticipated; and
   (ii) to charge for carriage by whatever method at the Company's option and packing; and

(iii) to charge for any additional design work or amendments to the specifications of the Goods used for quotation purposes or to the Customer's requirements carried out by or on behalf of the Company at the Customer's request; and
(iv) to charge for storage and other additional costs if the Customer fails to give delivery instructions within 14 (fourteen) days after the Goods are ready for despatch and, in such event, to invoice at the original delivery date.

3c All prices are ex-works unless otherwise stated by the Company in writing.

3d All prices are exclusive of value added tax which shall unless otherwise stated be payable in addition by the Customer.

4 *Delivery*

4a Dates for Delivery are given in good faith but are not guaranteed.

4b Risk shall pass to the Customer on Delivery, or in the event of delayed delivery pursuant to Clause 3 of the Conditions, from the commencement of the delay.

4c Goods must be signed for unexamined and the Customer shall notify any claim in writing to the Company:
  (i) for damage within 5 (five) working days after receipt of the Goods; and
  (ii) for shortage (subject to Clause 4d of the Conditions) within 10 (ten) working days after receipt of the Goods or invoice (as the case may be), and in all cases (other than non-Delivery) the Company must be given a reasonable opportunity to inspect the Goods in the state or condition in which they were received.

4d The Company reserves the right to deliver the Goods:
  (i) in more than one consignment and to invoice each consignment separately. Failure to deliver one consignment will not vitiate the Contract so far as it relates to the remaining deliveries.
  (ii) in quantities 10% above or below the scheduled quantities as may be considered reasonable having regard to the quantities ordered and the time interval between successive consignments.

4e In the event that the Customer shall delay delivery by more than 14 (fourteen) days by reason of its failure to supply essential parts by the due date, the Company reserves the right to invoice the Customer for all affected product at the agreed price for completed product.

5 *Payment*

5a Payment shall be made by the Customer without discount within 30 (thirty) days after receipt of the invoice.

5b Unless otherwise agreed by the Company in writing, a deposit of 50% of the quoted prices of all design, layout, artwork, mask and tooling charges is required with the Order. The remainder of all such charges is payable at the time of receipt of the first prototype or first production batch, whichever is applicable.

5c Time is of the essence in payment of all invoices and the Company reserves the right to suspend deliveries or to terminate the Contract when payment is overdue.

5d The Company reserves the right to charge interest on overdue accounts at the rate of 5% per annum over the Base Lending Rate for the time being of the . . . Bank.

6 *Representations, Conditions, Warranties and Exclusions*

6a The Company will within a reasonable time:
  (i) at its discretion replace or repair Goods found to its satisfaction to have been damaged in transit.
  (ii) credit and make good any admitted shortages at the previously invoiced price.
  (iii) deliver Goods found to its satisfaction not to have been delivered to the Customer.

6b The Company will at its discretion replace or repair free of charge any of the goods found to its satisfaction to be defective by reason of faulty materials or workmanship provided the Goods are returned to the Company carriage paid immediately the Customer becomes aware of the defect and in any event no later than 1 (one) year after Delivery and provided further that the Goods have not been tested, used, stored or maintained in a manner or for a purpose other than that for which they were specified, designed and tested, and provided further that the defect does not arise from components or other items supplied by the Customer or his agent or from components or other items specified as to type and manufacturer by the Customer or his agent.

6c In the case of Goods or their components or other parts not manufactured by the Company, its liability shall in no circumstances extend beyond the liability to the Company of the manufacturer of such Goods, components or other parts.

6d Where the Company has reason to believe that the testing applied by the Customer or his agent or by the Customer under the Customer's explicit instructions or the Customer's usage may be giving rise to defects and rejection by the Customer, the Customer shall afford the Company's personnel reasonable access to the workplaces in the premises on which the testing or usage is occurring in order to assist in the assessment of liability in a fair and reasonable manner and to permit suitable corrective action by both parties, as appropriate.

6e Unless otherwise agreed by the Company in writing, for custom circuits the Customer is responsible for ensuring that the component electrical values and tolerances thereon chosen for the design, and any mandatory or preferred component types and suppliers enable all the specified requirements for the completed circuit to be met with 100% yield when all such components perform to their respective specifications and are correctly connected. This condition also applies under any "worst case" circuit operating conditions, whether specified or experienced in normal use and with due allowance for parametric changes in performance over the life of the Goods which should have been allowed for in the electrical design. Any costs arising from failure by the Customer to comply with this Condition 6e will be for the Customer's account.

6f Goods returned hereunder will, if found not to be defective or if containing a defect for which the Customer is responsible, be returned to the Customer at its expense and subject to a handling charge of 10% of the invoice price of the returned Goods together with Value Added Tax thereon if applicable.

6g The Company's obligation to repair or replace the Goods is the sole liability of the Company (except in the case of death or personal injury caused by negligence within the meaning of Section I of the Unfair Contract Terms Act 1977) and all other representations, warranties, conditions, terms and statements, express or implied, statutory or otherwise, are hereby excluded.

6h The Company shall not be liable for any other direct consequential or other loss, or loss of profits (including but not limited to loss of data) of whatever kind and howsoever arising.

6i The Customer shall not rely on any representation concerning the Goods unless the same shall have been made by the Company in the Contract in writing.

6j Having regard to the ability of the Customer to obtain insurance cover in respect of the Goods, the liability of the Company shall in any event be limited to the invoiced price of the Goods.

6k Where free issue materials are supplied by the Customer, the Customer will be responsible at all times for loss or damage to these materials whether in transit to or from or at the Company's premises or the premises of the Company's subcontractor.

7 Title

7a Notwithstanding delivery of and passing of the risk in the Goods, ownership of all Goods supplied by the Company shall remain vested in the Company until payment for these Goods has been secured by the Company in full.

7b The Company is expressly authorised, in the event of the failure of the Customer to pay for the Goods by the due date, to enter on to the Customer's premises and retake possession of the Goods for which payment or part payment remains due.

7c Such payment will become due immediately upon the commencement of any act or proceeding in which the Customer's solvency is involved.

8 *Health and Safety*

8a The Company hereby gives notice that it has available information or product literature concerning health and safety aspects of such materials as are specified by the Company when used in the manner for which they have been designed and tested. Unless the Customer requests such information immediately on receipt of the Company's Order acknowledgement, it will be assumed that the necessary data are already in the Customer's possession and that the information and advice available from the Company is not required.

8b Where the Customer supplies components or other items for incorporation in the Goods, it shall ensure that sufficient information on the health and safety aspects of such products is made available to the Company to enable their use in a manner which minimises any associated risks and hazardous consequences.

8c The Customer shall be solely responsible for and shall keep the Company indemnified against any loss, liability or expense arising from use of the Goods other than in accordance with the Company's operating instructions or (where no such instructions exist) in a manner which could not reasonably be considered acceptably safe and with minimum risk.

9 *Patents*

The Customer will indemnify the Company against all actions, costs (including the costs of defending any legal proceedings), claims, damages or other expenses which may arise from alleged infringement of patents, trademarks, registered designs, copyright, intellectual property or other rights by Goods made to the Customer's designs or specifications.

10 *Specifications, Drawings, Photomasters, Computer Programs*

10a The Company will use all reasonable endeavours to maintain strict commercial security for all customer documents, drawings, photomasters and computer software entrusted to it.

10b All intellectual property rights in tools, jigs, dies, layouts, designs, manufacturing equipment and processes, inspection methods and procedures or any other matter made available by the Company to the

Customer in connection with the Goods shall remain the property of the Company and not be passed or disclosed to any third party without the Company's prior agreement in writing.

11 Force Majeure
The Company shall have the right to cancel or delay Delivery or to reduce the quantity delivered if it is prevented from or hindered in or delayed in Delivery of the Goods through any circumstances beyond its reasonable control, including but not limited to strikes, lock-outs, labour disputes, armed conflict, civil disturbance, or riot, fires, floods, acts of God, acts of Government, currency restrictions, unavailability or shortage of materials, breakdown of machinery, or failure of supplier, carrier or its subcontractor to deliver on time.

12 Default
If the Customer shall commit any breach of the Contract or be or become insolvent or unable to pay its debts or commit any act of bankruptcy or (being a limited company) go into liquidation other than a voluntary liquidation for the purpose of amalgamation or reconstruction only or have a receiver appointed of its undertaking or assets or a substantial part thereof or have any execution levied upon its goods or assets, the Company may without notice terminate the Contract or the unfulfilled part thereof and stop any Goods in transit without prejudice to any other right or remedy which the Company may lawfully enforce or exercise.

13 Cancellation
If the Customer cancels this Order the Customer shall pay to the Company all costs incurred by the Company in respect of this Order and 20% of the value of this Order.

14 Law
The Conditions and the Contract shall be governed by ... Law and the parties hereby submit to the non-exclusive jurisdiction of the ... Courts.

4

Assembly technology options, advantages and disadvantages

4.1 Introduction

A wide variety of printed board, hybrid and other assembly options are available, though in many instances the designer's choice is restricted. Typical restrictions come from circuit application category, cost factors, board/substrate dimensions, and the packaging styles available from semi-conductor and passive component manufacturers. Further limitations may be imposed by the size and weight of bulky items such as plugs, sockets and transformers.

The principal method of making electrical connection to component terminations—and, in most cases, also the mechanical fixing—remains soldering. Wire wrapping (cold welding), welding (parallel gap) and welding to posts inserted into the board exist as occasional options. Conductive adhesives are still lurking in the wings awaiting a wide range of components having compatible termination surface materials.

In selecting the most suitable assembly method and in deriving the best advantage from that choice, the electronic circuit designer must be aware of the advantages and disadvantages of each option and of the respective constraints that dominate results. Many of these are covered briefly in this chapter; for surface mount and hybrid circuits, more details are given in Chapters 5 and 6.

4.2 Through-hole (conventional) printed board assembly

Sometimes referred to as 'pin-in-hole' assembly, through-hole techniques have been the mainstay of electronic board construction over the past 50 years. Figures 4.1 and 4.2 illustrate, respectively, the basic style and a typical process sequence.

The techniques range from simple hand preparation of lead length and

ASSEMBLY TECHNOLOGY OPTIONS 49

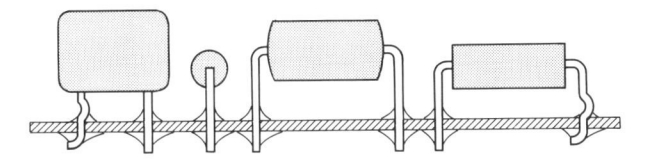

Fig. 4.1 Basic through-hole assembly structure

lead forming with hand insertion and soldering at around 100 components per hour, to sophisticated auto-insertion machines capable of cropping, forming and inserting the leads at rates exceeding 10 000 components per hour. Auto-inserters receive the appropriate mix of taped passive and discrete active components—which have been pre-loaded on to a single tape or bandolier by a sequencing machine—and via tubes or magazines containing integrated circuits. They are often linked mechanically to wave-soldering machines capable of matching their throughput capacity.

Many of the assemblies made for low-cost applications use the early type of single-sided board, i.e. with copper tracks and pads confined to the face opposite the components. Circuits made for the vast majority of professional users employ copper areas on both faces, with plated-through holes to interconnect outer and inner layers and which provide robust connections to component lead wires and tapes.

At time of writing, the cost of mounting through-hole components ranges from fully automatic handling at 1–2 pence sterling per component at very high volume, to 25–30 pence for hand-soldered items inspected to 'hi-rel' standards.

ADVANTAGES
- The comparative uniformity of the solder joints for differing components, but all having leads with standardized cross-section dimensions
- The ability to apply visual inspection to the joints when viewing from at or near 90 degrees to the plane of the board

Both of these make for ease of process control.

However, it has been traditional for the inspection to cover both sides of double-sided boards—just to check that surface tension has encouraged enough solder to wet both the lead and most of the hole.

- A further advantage from the long-term reliability standpoint is that the component bodies sit above the board during its passage through the wave-soldering machine. The temperature profile peak they see is usually below 130°C.
- The ability to apply electrical probes to the leads on the component side

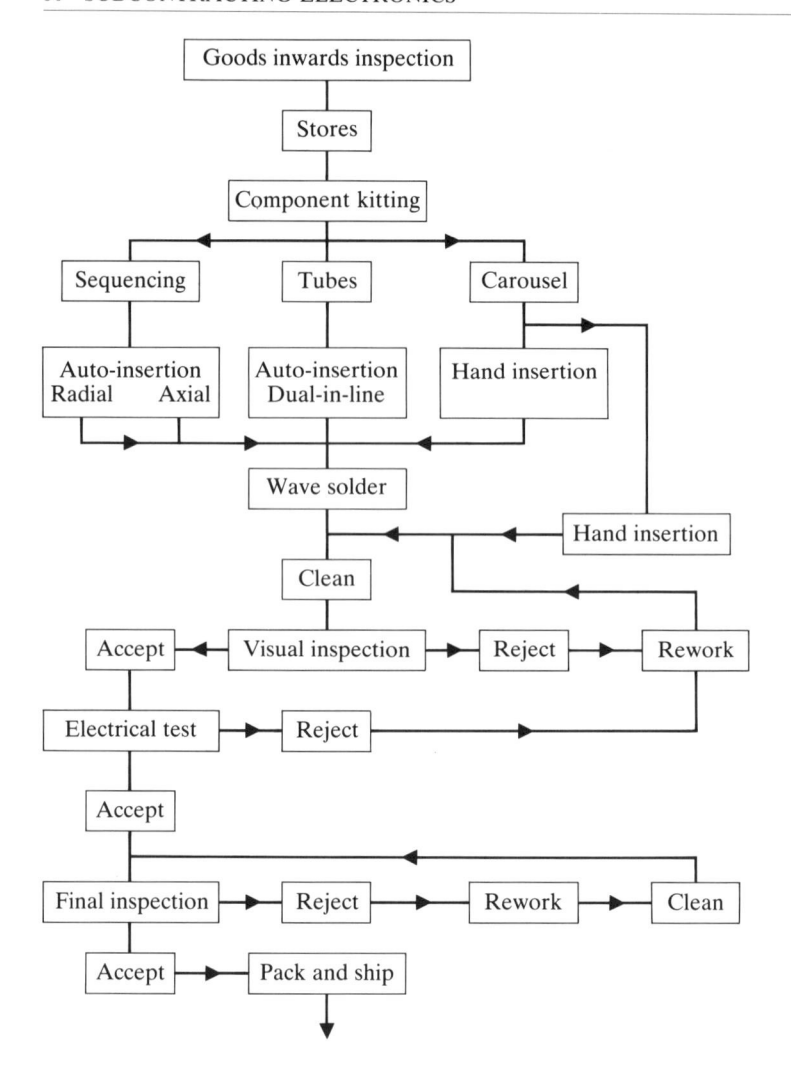

Fig. 4.2 Typical through-hole assembly process flow chart

of the board without prejudicing detection of dry joints is another plus point compared with surface mount assemblies.
- Because most of the components and their packages are well-matured products made by a large number of competing companies, the benefits of lower cost are passed on to purchasers.
- Power dissipation, in relation to both component package surfaces and the thermal conductance of comparatively substantial outgoing leads, has

minimized the need for forced cooling in a wide range of applications. The limits on component packing density have also contributed to this effect. This is not to say that forced cooling has been entirely eliminated, but merely to note that the mechanical limitation on the 'components-per-square-inch' figure, combined with the ability of semiconductor manufacturers progressively to reduce the power needed per transistor within integrated circuits, has enabled more through-hole assemblies to live without fan-assisted cooling.

DISADVANTAGES
- The comparatively large physical size and weight of through-hole components has limited their ability to provide highly miniaturized assemblies.
- If mass soldering techniques are to be used, the components can be mounted only on one side of the printed board—again restricting the ability to miniaturize economically.
- The emerging need for high pincount integrated circuits has brought packages with lead spacings well below the 25-year-old standard of 0.1 inch, for which through-hole mounting has become an impracticable method.
- In a few instances, the component body weight factor in relation to standard lead materials and cross-sections has raised difficulties. There has not been enough strength to resist stringent acceleration, shock and vibration specifications without adding support in the form of embedding encapsulation. The latter process can bring chemical compatibility and mechanical expansion effects which may harm reliability.
- As indicated earlier in this chapter, mechanized assembly may require the use of a sequencing unit to pre-load a bandolier with the correct range of component types before presenting them to the auto-insertion machine. This is an expensive piece of machinery.
- In itself, the insertion of component leads through holes is also disadvantageous in that one, two or three preceding mechanical operations have to be carried out, depending on package format. Radial leads require cropping to length, and may or may not need forming before entering to the board. Axial lead items need to be cropped to length, bent to the correct form and entered. In either case, an additional lead-joggling operation is sometimes used to provide a seating which holds the component body clear of the board. Where the board has to receive additional through-hole components inserted by hand before soldering, it is common practice to bend (clinch) the protruding ends of auto-inserted component leads to ensure that the components do not fall out and remain fully inserted during handling.

The above operations add complexity and hence both operating and tooling costs compared with the simple lifting and placement of surface mounted components.

- Inward pressure is applied to the rows of splayed tape leads of dual in-line integrated circuit packages to align their tips accurately with standard hole arrays. Once inserted, the outward spring pressure applied by these leads to the sides of the holes is intended to hold the component in place and thus avoid the need for clinching. Single in-line modules may be inserted without forming—if their lead tips are well aligned.
- The design of dual in-line semiconductor circuit packages and the need to force their leads inwards for insertion is believed to have brought additional mechanical stress and reliability problems into play. The reasoning behind this statement has emerged only since the arrival of surface mounting. The comparative reliability of surface mount equivalents in humid conditions, which for several sound scientific reasons should be much worse, in fact turns out to be considerably better.

4.3 Surface mount printed board assembly

The basic concept discussed here is to attach both leaded and leadless small components to copper pads (footprints) directly to the surface of the board. This means that the component bodies and their joints to the board are on the same surface. The opposite applies to through-hole assemblies. This technology has been available for over 50 years and until the 1980s has been applied mainly to thick and thin film circuits in which the substrates were glass or ceramic tiles. Figure 4.3 shows the surface mount structural options available. Figure 4.4 indicates process sequence options.

It should be noted that, although the majority of all surface mount assemblies use soldering as the joining material, there is a small but growing group of applications in which conducting organic adhesives may offer

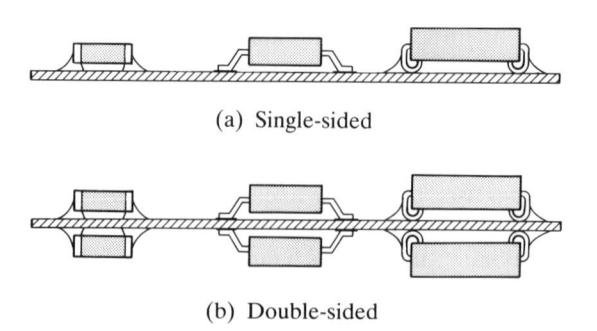

(a) Single-sided

(b) Double-sided

Fig. 4.3 Structural options for surface mount assemblies

ASSEMBLY TECHNOLOGY OPTIONS 53

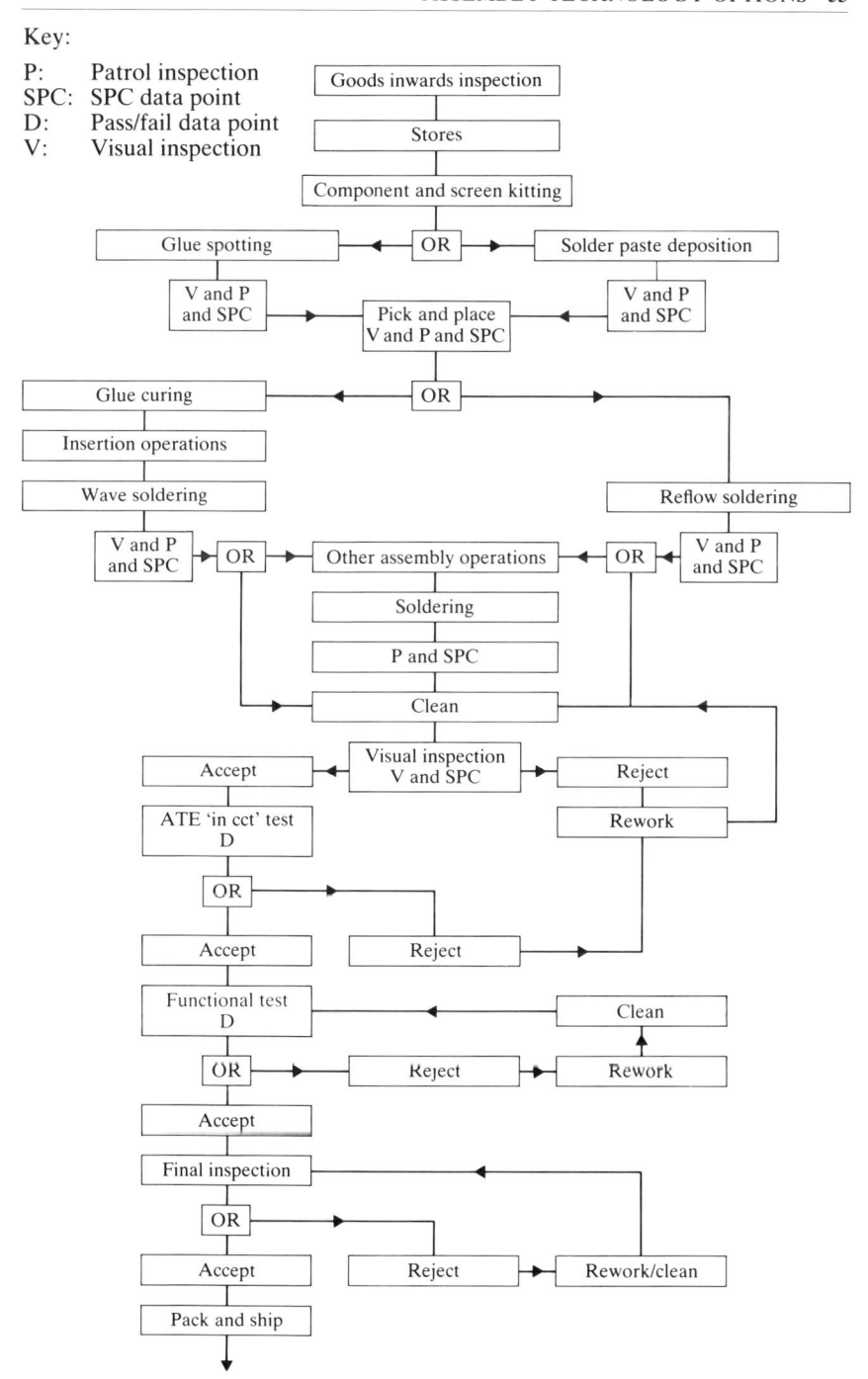

Key:

P: Patrol inspection
SPC: SPC data point
D: Pass/fail data point
V: Visual inspection

Fig. 4.4 Typical surface mount and mixed technology process sequence options

satisfactory solutions. However, for the use of adhesive assembly techniques to emerge as a major activity, it will probably be necessary to redesign the termination metallurgy of most components. Solder coatings are not suitable for the current range of adhesives.

Surface mount components come in two basic forms: leadless—often referred to as 'chips'—and leaded types. Apart from the change in asssembly structure involved, the leads and chip termination areas have disciplined fixed geometries. Unlike the through-hole situation, the dimensions of their board occupancy cannot be materially altered at the whim of the layout designer.

It was not until the end of the 1970s that surface mountable components began to be applied in production to epoxy–fibreglass printed boards. This sparked a sudden realization in the minds of component manufacturers that what they had previously perceived as packaging styles with a very limited market were, in reality, presenting them with a golden opportunity. The word 'golden' is applied with good reason, because it has enabled them to make huge economies in material usage while, for a considerable length of time, selling the same product functions at higher prices than their established conventional equivalents. In many instances that time-span includes the year 1992.

At the same time, it is argued that the lack of availability of a useful range of components was a major factor in delaying the uptake of the new technology. Capacitors, resistors and discrete semiconductor devices were the first types to appear, to be quickly followed by electrolytic capacitors, trimmer resistors, trimmer capacitors and low-pincount integrated circuits. Today there is a full range of passive and active components available, including inductors, relays, switches, light-emitting diodes, plugs and sockets and medium-power transistors.

In some applications, the deposition and curing of semi-organic thick film resistor materials directly on to the surface of the printed board can avoid the need for added component equivalents and has been used successfully. Their ohmic value tolerance and stability are inferior to normal inorganic thick film types, and they have been used only in situations where these parameters are not critical. There is also the question whether they may not react well when subjected to subsequent surface mount assembly processes such as soldering.

Semiconductor integrated circuit packages with leads emanating from all four edges are unique to surface mounting and do not appear in any through-hole format. The nearest to being an exception is the pin grid array in which 50–150 through-hole pins emerge from the large under-surface of a moulded circuit module on a 0.1 inch pitch. The pin grid array is not a favourite with OEM production engineers because of the difficulties in obtaining good post-wave-soldering yields and in carrying out reliable rework.

The comparative simplicity of surface mount assembly operations has meant that mechanization came very quickly for thick film circuit assembly, and it was machines developed for making hybrid circuits that formed the basis of today's high-speed component pick-and-place equipment for printed boards. Their rates of placement currently range between 1000 and 17 000 per hour; by ganging several machines together, up to 250 000 per hour is achievable.

The price of these machines is often expressed in 'pounds per placement per hour'. Thus, a machine costing £100 000 and placing 3333 components per hour would be rated at £30 per placement per hour. This was a typical figure for 1986–8, but today the number of machine manufacturers has extended tenfold and competition has meant that figures between £15 and £25 are commonplace.

Instead of bandoliers, components are presented in tape on reels, in tubes (tubular magazines), in flat so-called 'eggbox' type trays and, for earlier machines, in bulk from vibrated hoppers or small cartridges.

However, unlike through-hole component assembly, reliable products based on hand assembly in surface mounting are only marginally practicable, with even greater skill being required as chip passives get progressively smaller, e.g. 1 mm × 0.5 mm. Also, we have multi-lead packges extending their pincount to several hundreds with lead pitches descending below 0.5 mm (20 mil) to 0.1 mm (4 mil). Reliability, in this case, requires strict control over maximum temperatures and especially a very close control of component heating and cooling rates in the soldering process. In addition to the excellent manual dexterity needed, background knowledge and intelligence are required when attempting hand soldering of SM assemblies. Undetectable damage causing subsequent field failure is the penalty for ignoring these requirements.

However, there is no reason why prototypes should not be assembled by hand, although the safe advice is that they should not be shipped to customers other than for circuit function verification and, possibly, for thermal assessment. Some of the reasons for the above statements are amplified in Chapter 18.

Nor should they be used in type approval testing—particularly for hi-rel applications or where use over a wide temperature range is intended. The level of intermetallic formation cannot be sufficiently well controlled during hand soldering, and this affects both joint strength and the ability to withstand regular thermal cycling.

Allied to mechanized component handling has been the introduction of new reflow mass soldering techniques. Typically, these include the use of infra-red radiation, condensation (vapour phase) and hot gas flow as heat sources for reflowing solder paste mounds. These mounds have to be deposited immediately before placement so that they appear below the component terminations and the 'tackiness' of the paste is sufficient to hold the components in position while the board is transferred to the soldering equipment. Methods of depositing the solder paste include screen (off-contact) printing, stencil (in-contact) printing, dipping the component terminations *en passant* during the act of placement and, finally, dispensing from a syringe. The vast majority of users apply one of the two printing techniques.

For a variety of technical reasons, the component packing density on the board using reflow methods can be much greater than for wave (flow) soldering. To use the latter process, surface mounted components must first be glued to the under-side of the board before it is passed through the flux applicator and then through the wave in a flow-soldering machine.

The first mass production boards using surface mounting came in Japanese consumer products. The termination areas of standard miniature $1/8$ watt tubular resistors with their leads omitted (now known as 'metal electrode face' or MELF components) were soldered directly to the board. Usually they were single-sided assemblies.

ADVANTAGES
- The obvious advantages of reduced size and weight have been seen in many countries as the most important features of surface mount technology, especially where the electronics industry is spread across the whole public user spectrum, as in the USA, the UK and France.
- Where very high-volume production of consumer items predominates, for example in Japan, the miniaturization factor is not pushed to its limits because lower production costs are more important. These are achieved through the comparative ease of mechanization and automation of the complete surface mount assembly process.
- Semiconductor component costs are, on average, still above those of through-hole equivalents, but on a pence-per-function basis they are steadily reducing.
- the smaller size and closer proximity of components brings reduced interconnection path lengths, leading to lower parasitic impedances and faster system operating speeds in digital circuitry.
- The smaller size results in less sensitivity to external electromagnetic interference and a reduced risk of transmission of the same effects.
- For the same functions, the overall equipment box size is reduced, leading to lower transport and storage costs, especially for exports.

DISADVANTAGES
- The increased number of variables in design and manufacture demands

staff of a higher calibre and increased training costs at all levels. Failure to achieve the required degree of professionalism can be very costly.
- Smaller gaps between tracks on printed board layouts can increase the risk of lower inter-track impedances, allowing spurious noise and crosstalk.
- The same small gaps may make it necessary to provide low free ion-content surface varnishes to give protection against adverse external environmental conditions. Very thorough cleaning is needed prior to varnishing, otherwise the existence of trapped ionic contamination beneath the layer can cause even more severe corrosion.
- The capital investment needed to achieve reliable products is high. The penalties for purchasing inferior soldering equipment may also be high (cf. Section 17.2).

APPLICATIONS
The proportion of circuits that can be made containing only surface mount components is still quite small in comparison with mixed-technology types. However, in many instances the number of through-hole components is minimal and they can be inserted and soldered by hand without adding much cost.

This last point has pushed the application of surface mount assembly into almost every corner of the electronics scene.

4.4 Mixed-technology printed board assembly

This approach is more than mere pragmatism. Component availability, cost and assembly ruggedness have combined to ensure a current OEM regime of mixed technology in which both through-hole and surface mount components are combined on the same board.

Because most through-hole components are not designed to withstand the high soldering temperatures applied to surface mounted components, the techniques used for mixed-technology assemblies must include wave and/or hand-soldering. The range of structures available and the relevant process sequences are shown in Figs. 4.4 and 4.5.

ADVANTAGES
- The electronic circuit design can be optimized to take benefit from a favourable combination of lower component costs.
- The use of conventional through-hole items is often an important factor in achieving high reliability—especially for plugs and sockets and other heavier components. For these types, the adhesion strength of surface

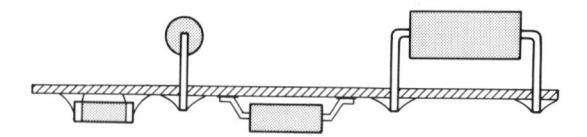

(a) Double-sided mixed technology, components on opposite sides

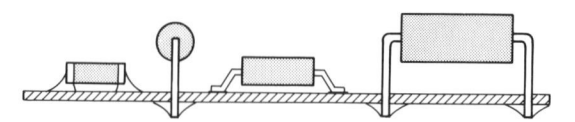

(b) Single-sided mixed technology, all components on same side

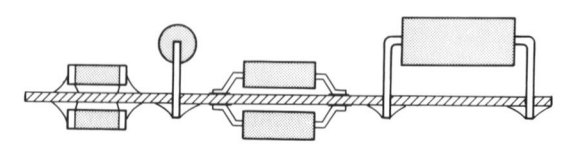

(c) Double-sided mixed technology, SM components on both sides, TH components on one side

Fig. 4.5 Structural options for mixed technology

mount pads and solder joints may not always be sufficient to cope with shock, vibration, wide temperature ranges and, above all, mishandling. In a few projects, larger ceramic chip capacitors have been replaced by radial lead equivalents to avoid stress failures arising from TCE differentials between board and component.
- Mixed-technology assemblies can be made with a single pass through a wave-soldering machine.

DISADVANTAGES
- Where a single-pass wave-soldering assembly method is preferred, the surface mount component packing density on the underside of the printed board is lower than for reflow soldering. Wider spacings are needed to avoid unmade joints or 'skips' which can occur when terminations are shadowed from the wave by adjacent component bodies. On longer-term reliability grounds, some users prefer to avoid the thermal shock generated by passing semiconductor plastic packages through the solder wave.

ASSEMBLY TECHNOLOGY OPTIONS 59

APPLICATIONS
Because this method offers a good combination of reduced process costs and lower component prices, it is most widely applied in high-volume applications, including consumer and automotive products.

4.5 Hybrid (film) circuits

Hybrid film circuits have been in use for over 50 years. Typical structures are shown in Fig. 4.6.

The word 'hybrid' is hard to define because it is often applied to a wide variety of circuit construction forms and also to computers having mixed serial and parallel processing. In this context it used to describe the range of

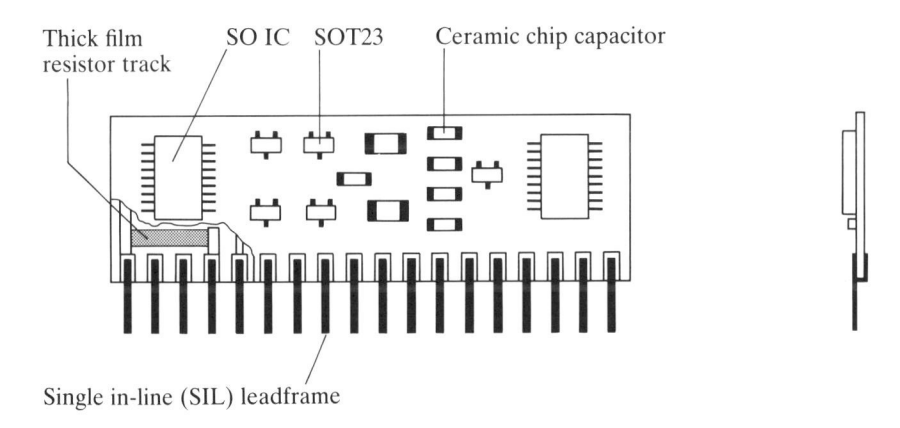

(a) Thick film SIL hybrid

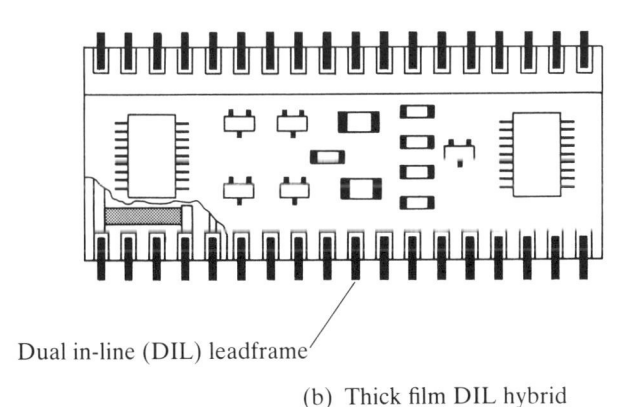

(b) Thick film DIL hybrid

Fig. 4.6 Typical thick film hybrid structures

assemblies based on substrates, usually having ceramic or glass surfaces, to which combinations of conductive, resistive, insulating, dielectric or semi-conducting films have been patterned and made adherent. The later addition of surface mounted components and leads by soldering, bonding or use of adhesives completes the assembly, and in some applications encapsulation follows.

Hybrid circuits containing no semiconductors are described as 'passive networks', irrespective of whether the components are surface mounted or exist as deposited films on the substrate.

The temperatures at which the films are made integral with a ceramic surface vary from 'curing' or adhesives at 100 °–175 °C to 'firing' of glazes in a furnace (kiln) at 500 °–1000 °C, depending on whether the films are organic, inorganic or a mixture of both.

When the films are deposited and fired-on as glazes, the structures are known as 'thick film' circuits. When they are deposited in vacuum or by plating, they are usually described as 'thin film' circuits. Both exist on ceramic and glass substrates.

Versions of these structures which use organic printed board substrates are now known as surface mounted assemblies, and these have reduced dramatically the market areas open to ceramic-based hybrid equivalents. The reasons include cost, ease of design and production in double-sided and multilayer format and, above all, economic miniaturization.

ADVANTAGES

The niche market options for ceramic-based hybrids include:

- *High temperature applications* Ceramic substrates with fired-on films and added components made in ceramic format—e.g. leadless ceramic chip carriers for discrete and integrated circuit semiconductors, ceramic chip capacitors—and fired-on film components are capable of operation up to 150 °C.
- *Power dissipating circuitry* Ceramics such as alumina are good heat conductors compared with organic printed board materials. They form excellent substrates for circuits containing power transistors and also for very high-density, high-speed logic arrays in military and avionic applications.
- *Hermetic encapsulation* Ceramic substrate structures are widely used to create dry gas-filled or 'cavity' packages in which the bond wires from semiconductors to the surface films are not highly stressed or contaminated by surrounding encapsulation materials. Wire bond strengths at the lower end of the distribution spectrum can survive temperature cycling for longer periods than those embedded in moulded or cast plastic materials.

- *Microwave applications* Although alumina has a higher dielectric constant than many PTFE-based printed board materials, its stability in this parameter is much better over a wide temperature and humidity range. Borosilicate glass substrates have some advantage because of their low dielectric constant. For this reason, some microwave hybrid circuits are made using thin film technology in which metals are deposited in vacuum chambers. The accuracy of subsequent photo-chemical etching to form the conductive patterns is better than for thick films on ceramic, but the films are thinner, and the higher ohmic resistance of conductive tracks may require extra plating-up processes.
- *Long life equipment* Ceramic and glass substrates and deposited films can be manufactured with very low ionic contamination levels. This means that the parametric stability of precision film and other miniaturized components can be maintained over many years. Circuits that are comparatively inaccessible for repair, for example in communication satellites and under-sea telephone cable repeaters, can benefit from this advantage.
- *Dynamic trimming after assembly* The use of lasers and abrasive jets to adjust the value of deposited resistors after assembly while electrical test is being carried out is called 'dynamic functional trimming'. This is an extremely useful feature because it allows greater precision in functional performance and the use of some lower-cost wide-tolerance components, for example in precision filters. This technique can avoid the need for 'select on test' components. It is not unique to film circuits, but is much easier to apply when the substrates are inorganic.

DISADVANTAGES

It must be said that, while the disadvantages are few in number compared with the advantages, they have been predominant in causing the market shift towards printed board equivalents.

- *Control limitations* Successful manufacture of hybrid film circuits requres a control over materials, processes and factory production environments that is more akin to the disciplines used in semiconductor manufacture than to those seen in printed board assembly areas. This adds to capital, operating and other investment costs. A corollary of this is that the market perception of hybrid circuits is that they are more likely to be subcontracted to specialist companies. Printed board SM assemblies are seen as suitable for in-house manufacture by OEMs, thus retaining their product's added value content.
- *Size limitations* The brittle nature of ceramic and glass substrate bodies and surfaces imposes severe size limitations on the range of practical

applications. For this reason, they are often seen as large added components on mixed-technology surface mount or conventional through-hole board assemblies.

- *Multilayer circuits* There are production yield and hence cost problems which inhibit the use of printed and fired multiple conducting/insulating layers on most ceramic-based thick film hybrids. The same restriction applies to the provision of through-hole connections to join conductor tracks on opposite faces of the substrate. This has limited the cost effectiveness of hybrids in the miniaturization field for professional and consumer applications and has also inhibited the successful mounting of conventional through-hole components on ceramic.

For very high-quantity production, some of the above difficulties can be avoided by carrying out the multilayering and interlayer connections as part of the original substrate manufacture while the ceramic material is still in the soft clay (green) state, i.e. before it is fired as a flat slab in the kiln furnace.

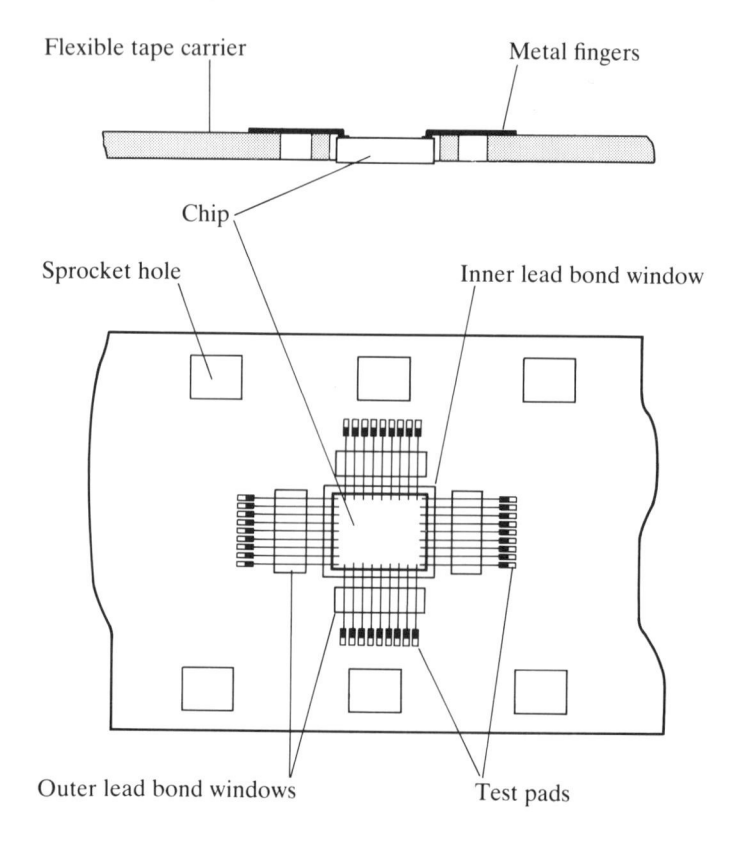

Fig. 4.7 Tape automated bond (TAB) device structure

4.6 Tape automated bonding (TAB)

This method of presenting semiconductor devices for mounting to printed boards and other substrates is one of the smallest-size options available to date (cf. Fig. 4.7).

The TAB device consists of an intermediary thin printed board carrier to which the semiconductor die is mounted—either by conventional methods or using a conducting adhesive. Connections between the outgoing pads on the die and the metal tracks on the carrier are either via overhanging tape leads fabricated integrally with the semiconductor die, via similar leads made integrally with the printed board carrier, or via classical wire bonding methods.

In the second case, the carrier material is usually a high-quality epoxy or 'Kevlar' film without fibreglass reinforcement. The omission of the latter is primarily to facilitate the etching process, which enables the base material to be removed selectively and allows the overhang of the metal prongs that make the electrical connections—inward to the chip and outward to the printed board.

The preparation of both chip and carrier requires careful control of the surfaces to be bonded. Where soldering is the preferred method of attaching the inward leads to the chip, a solder mound (bump) on the lead (or on the chip itself) is needed. Typically, this is of the order of 20–30 microns high and of a diameter to suit the relevant pad area on the chip, e.g. 80–100 microns. Control of the final collapsed thickness of the solder joint is necessary to avoid damage to the chip surface.

Because semiconductor manufacturers are reluctant to supply 'bumped' chips in small quantities, many users opt to go for 'bumped' leads—even though the number of good carrier suppliers is not great. Alternatively, some silicon foundries will supply tabbed chips ready for testing and subsequent auto-attachment to organic printed boards or ceramic substrates. These are basically non-integral beam-lead devices in carriers.

ADVANTAGES
- The electrical testability of TAB devices before mounting is considerably better than for bare dice. In most cases, special test pads are provided to avoid contamination or damage to the leads which must ultimately be connected to the substrate.
- The additional flexibility given by provision of leads enables TAB devices to be mounted on a wide range of substrate material types without concern over differential thermal expansion problems.
- Their small board area occupancy enables TAB devices to offer cost-effective miniaturization in applications where small size is a dominant selling factor in the market-place, for example in mobile telephones.

DISADVANTAGES

- The availability of TAB-mounted semiconductors is restricted and often depends on their usage by a large customer or series of projects worldwide. In many cases a special chip layout is required.
- The costs in preparing the die and carrier and of joining the two together are usually higher for TAB-mounted devices than for plastic-encapsulated equivalents. Special 'bumps' are required, either on the die itself or on the carrier inner leads.
- OEMs attempting in-house assembly of dice to carriers and carriers to printed boards need hi-tech skills and equipment.
- The process of attaching the TAB device to a printed board normally involves a thermode bonder and cannot yet be carried out simultaneously with the mass-soldering techniques used for other components on the printed board. TAB-mounted devices are assembled after mass soldering.
- To achieve good thermal contact with the board, the addition of conductive adhesives may be required beneath the die.
- Suppliers of TAB carriers are restricted in number.

4.7 'Chip on board' assemblies

In this type of assembly, semiconductor dice are mounted directly to conductor areas on the printed board, ceramic or glass substrate with no intermediary carrier. After assembly and test, it is usual to apply hermetic or local resin encapsulations (globs) over the dice to protect them against humidity and mechanical damage.

Figure 4.8 shows three alternative methods of mounting. The first involves normal face-up attachment of the die using one of several alternatives—the standard eutectic method, a solder preform or a conducting adhesive. The operation is followed by wire bonding the connections from the die surface to the printed board, again using conventional techniques which may include heat and ultrasonics (thermosonics).

Where gold wire is used for these outgoing connections, it is usual to plate the receiving copper areas on the board with gold very slowly to obtain a malleable consistency and thickness suitable for successful bonding. The method of supporting the board or substrate below the bonding area is also critical to correct ultrasonic energy transfer.

The second method, illustrated in Fig. 4.8(b), is known as the 'beam lead' technique. The leads are deposited on the die while it is still part of the silicon slice and the area of silicon under the fingers is etched away to separate the dice and leave the 'beams' extending outwards beyond the edges of the die. The fingers remain integral with the die and their outer ends

ASSEMBLY TECHNOLOGY OPTIONS 65

(a) Face-up die

(b) Beam lead

(c) Flip chip

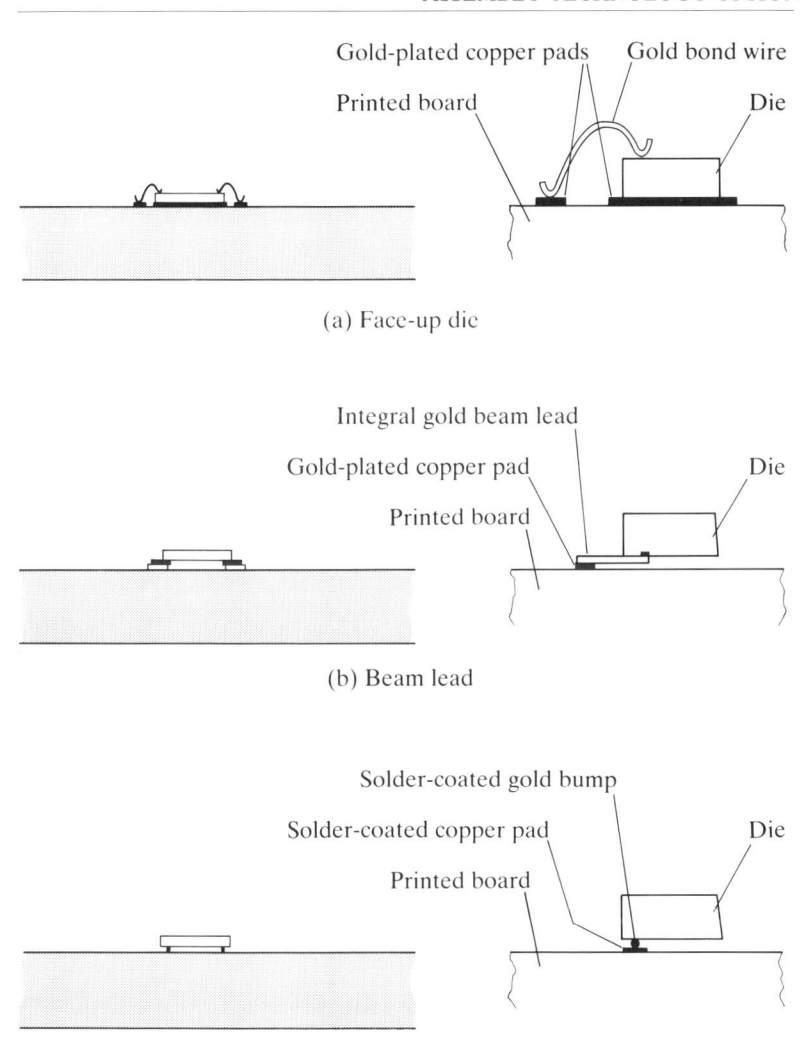

Fig. 4.8 'Chip on board' (COB) assembly types

are thermo-compression-bonded in a single operation, for example direct to gold-plated fine line tracks on the printed board.

The method of attaching beam-lead devices to the substrate is similar in concept to that used for TAB structures, but as no intermediary carrier is involved it is therefore classed as a 'chip-on-board' device.

The third method, illustrated in Fig. 4.8(c), is to invert the die and use reflowed solder or solderable pillars between matching pad areas on the under-surface of the die and on the board. This is known as 'flip chip' assembly and it requires the use of special techniques to obtain 'controlled collapse' of the solder to ensure a small clearance under the die after soldering. In some instances the solder has been replaced by a mound of other materials, e.g. soft gold, which can be thermo-sonically bonded direct to gold-plated pads on the board.

ADVANTAGE
- These methods achieve the greatest degree of miniaturization for hybrid circuits and printed board assemblies.

DISADVANTAGES
- Many types of silicon die cannot be fully tested electrically and thermally before mounting. This can lead to assembly yield problems and costly rework in replacing faulty dice—a factor that restricts the number of dice applied to a board.
- Normally the semiconductors must be mounted on the board after assembly of any components that have to be mass soldered.

4.8 Multichip modules (MCMs)

MCMs are merely specially packaged small 'chip-on-board' or 'chip hybrid' structures. They consist of an interconnective substrate, which may be of silicon, ceramic (e.g. alumina) or even a small printed board, on which are mounted two or more semiconductor dice; in some cases where resistors or capacitors are not integral with the substrate, other passive components in chip format, such as ceramic multilayer capacitors and resistors, are attached (cf. to Fig. 4.9).

The dice are attached to the substrate by classical silicon–gold eutectic bonding methods or using solder or Stage B epoxy adhesive preforms. Normal wire bonding can connect the outgoing pads on the silicon to those on the substrate. Added passive chip items are connected by wire bonding or soldering, or are mounted on the substrate with conducting adhesive, depending on their structure and termination materials.

The completed modules appear in a variety of package styles including through-hole leaded modules and surface mounted leadless ceramic chip carriers (LCCCs), flatpacks, moulded and cavity plastic quadpacks, printed circuit-type leadless cavity packages (EPICs) and, more recently, quad packages with fine pitch leads. The latest have leads emerging from four edges with spacings down to 0.3 mm. For this latter style, one can predict even smaller lead pitches in future.

Wire-bonded die   Interconnective substrate
                  (e.g. silicon or alumina)

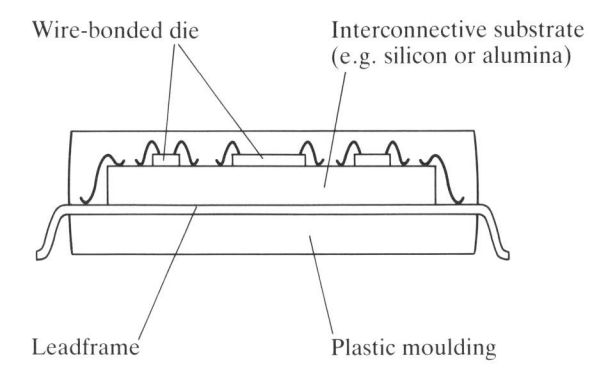

Leadframe   Plastic moulding

Fig. 4.9 Typical plastic-encapsulated multichip module

Basically, MCMs constitute just one more component that can be mounted on a printed board. The ceramic-based types have been around for more than 20 years under another name—multi-chip hybrids. These have been used widely in military, avionic and spacecraft systems where the cost premium was less important than the miniaturization factor.

ADVANTAGES AND DISADVANTAGES

Among the reasons why they have not taken off earlier in the market-place have been the difficult commercial interfaces and testing problems that occur when unmounted (bare) semiconductor dice are purchased by OEMs.

As mentioned above, it has been impossible to check completely the functionality and thermal properties of bare dice with enough accuracy. When, after mounting by the OEM, their performance is unacceptable, the question arises as to who caused the defect? Was the die itself faulty, or was the defect caused by the OEM when mounting it? Resulting assembly yield losses in mounting bare dice have limited the number of semiconductors that can be used cost-effectively on a single substrate.

A way around this problem has been the use of TAB devices in MCMs instead of bare dice. As indicated above, they can be tested more thoroughly, but do not enable quite the same degree of miniaturization to be achieved.

The other factor inhibiting the use of silicon dice in quantity by OEMs has been the reluctance of semiconductor manufacturers to lose added value by selling their dice at their prime cost plus a reasonable mark-up. This has been understandable because a large proportion of the capital equipment used by semiconductor houses is tied up in mounting, encapsulation and final electrical test equipment.

In a capital-intensive business there has to be a satisfactory return on this investment. Hence, although one could purchase, say, 100 dice at a good price, when you asked for 100 000 the answer was 'You will have to pay the same as for the encapsulated version.'

More recently, the ability to test the bare die has improved and also there are semiconductor companies who are geared up to sell only unmounted, unencapsulated dice. This has eased the situation for OEMs, but it is still evident that in many consumer areas where MCMs are now being used, the volume applications will be dealt with most competitively by the semiconductor suppliers themselves rather than hybrid manufacturers or electronic equipment OEMs. In these circumstances, the bare dice do not have to be sold across the above-mentioned difficult commercial interface.

APPLICATIONS

The first high-volume applications have been in mobile telephones, but it now looks as if their use will extend even into automotive electronics, where price and reliability are the major issues rather than small size.

5
Design checks before subcontracting SM assemblies

5.1 Background comments

Despite the advent of computer aided design (CAD) systems and other aids, when measured against the real need, the standard of overall design knowledge in the electronic industry has probably declined over the past decade. This appears to have been caused by a lack of awareness at very senior management levels.

Put simply, rapidly increasing circuit complexity and lack of early feedback via industry top management to universities and technical colleges has resulted in the failure of academe to provide suitable multi-disciplinary training for electronics engineers.

'Design for manufacture' is the driving force in this context. For 20 years or more it has been no longer acceptable to develop an electronic circuit or system design without concurrent evaluation of the detailed means of its creation. Yet the idea of 'leaving the production engineers to sort out the right technology'—a concept that should have been consigned to the dustbin long ago—is still widely practised and is responsible for many of the difficulties encountered by both sides in the subcontract business.

In addition to the principles of electronic circuitry, a broad standard of near-degree-level knowledge in physics, chemistry, plastics materials, metallurgy and mechanical engineering is now an essential baseline for electronics designers applying present and future interconnection technologies.

The power of CAD systems to help in this problem is enormous. So far the efforts of the electronic CAD software specialists have been channelled more towards the manipulation of gates, holes and electrons and two-dimensional substrates rather than towards the integration of such things as the third (z)-dimension component factors in printed board layout, thermal warnings, mechanical stress analysis and process sequence and planning factors.

Much of the software needed to give the missing support is available from work in other areas. It is the lack of multi-disciplinary thinking that has kept things apart when they could easily have been put together.

A second adverse result of the introduction of CAD in the UK has been a trend towards even further isolation of the design function from daily life on the assembly shop-floor. This has occurred partly through increased use of external CAD bureaux, partly from the mystique surrounding 'computer rooms', and partly in line with long-established large-company 'class' traditions which separated 'staff' from 'workers'. Fortunately the latter concept is gradually fading; indeed, it will have been eradicated long ago by competent management teams in small and medium-sized electronics companies.

Traditionally in larger companies there has been a separation between the 'Design Department', usually staffed by senior engineers, and the 'Drawing Office', normally staffed by draftsmen. Also traditionally, it has been the draftsmen who were responsible for doing the printed board layouts. The advent of CAD machines involved major retraining programmes, particularly for the draftsmen.

With the advent of surface mounting, despite an obvious need for very high intellectual calibre and a much wider breadth of scientific knowledge in the circuit design and board layout functions, the draftsmen were expected to carry on as before, with little or no additional training in the new requirements. After all, the management thought, surface mounting merely required a new set of production equipment and processes, and it seemed that suitable pad sizes on the printed boards were on offer from the component suppliers. Nothing to it!

Many companies were slow to recognize this mistake and persisted in giving the new work to their draftsmen instead of encouraging their best engineers to take over the task.

There are many examples substantiating the above statements, some of which are given in the case-histories in Chapter 20.

5.2 Electronic circuit design

The combination of miniaturization and mechanized assembly has brought new pressures upon the electronic circuit designer. The words 'right first time' embody the change.

The market, via the project manager, sets the pace by invoking the doctrine of maximum function in minimum volume. Closely sited components with negligible spare space on the printed board permit no easy additions.

Modifications due to circuit design errors usually require major re-layout and in many cases bring either an increase in manufacturing difficulty or an increase in board size, or both. Besides adding costs and time delay, the resulting hiatus can make the difference between success and failure in a highly competitive market-place.

For these reasons, the wise subcontractor will want to assess the calibre of the electronic circuit design before accepting an order—if only to get a feel for the likelihood of late modifications and their possible adverse effect on production schedules and cash flow. Typical questions on the subject which a customer should expect from competent professional subcontractors are given in Chapters 12 and 13.

5.3 Component selection

It is wise to check that the components selected by the designer are suitable for the subcontractor's intended assembly processes, particularly placement, soldering and cleaning. Reference to the manufacturer's data sheets is necessary because many suppliers nowadays are imposing severe process and rework constraints in order to minimize their product liability exposure.

Components that are too large, or too heavy, or cannot be manipulated with sufficient accuracy by the subcontractor's available pick-and-place machines, must either be hand-assembled or be made the subject of special mechanization—both of which can add to assembly cost. Where board cleaning after assembly is essential, the use of ultrasonics may be required to ensure solvent flow beneath chips and large integrated circuit packages. Not all components are suitable for such treatment in all liquids.

Circuit designers and buyers should be aware that apparent savings made by the purchase of cheap components may easily be more than offset by higher assembly costs. For example, the prices of metal electrode face (MELF) resistors are still lower than chip types, but higher rework, recleaning and retesting costs can result from their use in reflow-soldering situations.

5.4 Printed board specification and layout

Surface mounting has brought new imperatives to the circuit designer. These include a far greater knowledge of the materials and methods of board manufacture than was hitherto necessary. The list of parameters that the professional will need to consider in relation to the intended assembly process—and then to specify to the board manufacturer—has increased almost tenfold. It is important that the buyer is aware of the key items. Most of them are covered in the checklist given in Chapter 8.

In some senses there is now also a new role for the buyers, who, in the interests of their companies' projects, should see themselves as the board manufacturer's representatives. In this connection, there are two important tasks. One is to ensure that the order gives sufficient information to allow the right product to be delivered. The second is to ensure that the layout engineer is aware of the factors that can keep the board price to a minimum. Both may seem to be fatuous statements until the depth of knowledge required is understood.

For example, board costs can be reduced by arranging that the panel dimensions chosen can be at or near an exact fraction of the board manufacturer's untreated sheet size, e.g. 6 ft by 4 ft. This reduces the amout of waste material. Lack of appreciation of this simple opportunity forces most board manufacturers to operate a sheet usage efficiency of around 80 per cent.

The aspect ratio of the panel and its orientation on the pick-and-place machine platen can affect the component placement rate of some single-head equipments. Within the constraints of practical board shape and the step-and-repeat pattern required (if any), the dimensions should be aimed at enabling placement at or near the maximum rate.

More realistic standardization in component bodies and terminations could help in reducing some of today's soldering yield problems. Pad geometries should be tailored to the component outline dimensions given in the specified supplier's data sheets rather than to the issued standards. This does not necessarily mean using the component supplier's recommended pad sizes—if given. The need to apply the principle is more likely to occur with high-pincount packages from Far Eastern sources than with those conforming to JEDEC outlines.

When placement accuracy requirements demand optical alignment and automatic correction facilities in the solder paste printer, adhesive dispenser and/or pick-and-place machine, the board layout engineer will need to know the precise mark shapes, sizes and surface finishes that each of the subcontractor's equipments is able to recognize.

The subcontractor should be asked to specify the bare board flatness requirements needed for those processes that are sensitive to this parameter, and the information should be passed on to both the board layout designer and the board manufacturer. The operations affected are likely to include screen-printing of solder paste or adhesive, adhesive dispensing from syringes and automatic component placement. In some instances, the electrical test station may be added to the list, but this requirement is for the post-soldering situation.

Board manufacturers should be asked to supply guidelines for board layout and layer thickness. Together with their own good laminating practices, these will enable their customers' design engineers to help in achieving the best flatness. References on this subject are given in the Bibliography.

5.5 Large printed board assemblies

It is common today to find multilayer mixed-technology printed board assemblies up to 30 cm and even 40 cm square with surface mount components closely spaced on both board faces as well as conventional through-hole mounted parts. Such structures may carry more than 1000 components and have pincounts exceeding 4000. Examples are seen in the context of computer production, where all the electronic functions are assembled on a single board.

Simple mathematics dictates that, even if a reasonably good post-soldering joint defect rate of 500 parts per million is achieved, there will be an average of 8 defective connections on each board and the first-time yield of board on electrical test will be near zero. In these circumstances it becomes absolutely essential to design the layout with test points at every node in the circuit, so that all the components can be accessed individually for checking. This is called 'in-circuit' testing and may require from 1 to 15 per cent extra board area—a penalty well worth paying in return for the instant fault diagnosis capability it offers.

The electrical test is carried out using a matrix of probes or 'bed of nails' and the equipment acts more as manufacturing defect analyser than as a component performance checker.

Without such a test it is almost impossible to locate the individual dry joints which are often so hard to find visually on large boards. There can be hundreds of different soldered connection types having different shapes and sizes, and many are difficult to see, even under magnification. Even with the combination of in-circuit testing and visual inspection methods, a few dry joints may still escape early detection. Some companies use vibration during testing to find these.

Functional testing does not provide sufficient diagnostic power to pinpoint individual component joints. However, if the in-circuit test is passed, then the probability that the full functional test will also be passed is high.

Failure to follow the above advice on designing in in-circuit testing has been a key factor bringing bankruptcy to at least one computer manufacturer and has caused heavy financial losses for many companies arising from manual fault diagnosis and lengthy rework operations.

Often these problems come from a lack of management awareness and from ill-advised acceptance by project managers of excessive demands for miniaturization. For each assembly technology it is possible to establish sensible, cost-effective limits of size reduction. To step outside these limits invites a sharp increase in technical risk and time delay that can easily bring unnecessary financial hazards to the project. Figure 5.1 illustrates the principle.

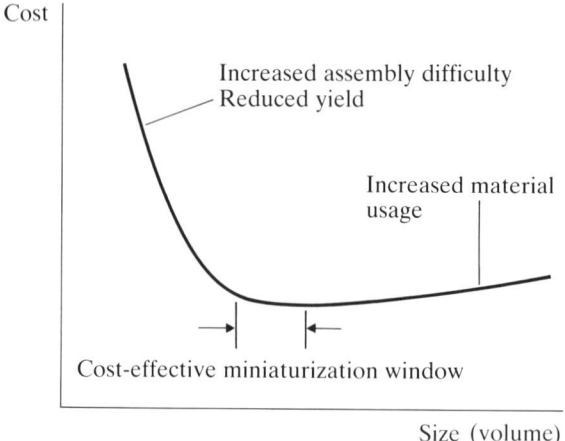

Fig. 5.1 Cost–size relationship for assembly technologies

5.6 Equipment structure

Many excellent board layouts have been rendered unreliable though failure to understand the need to apply simple stress analysis to the structural situation of the board in the completed electronic hardware system. The reality is that the intended method of entering or fixing assembled surface mount and mixed-technology boards into (and out of) the equipment requires design attention before layout commences.

The objective is to ensure that sensitive surface mounted component bodies (e.g. thin ceramic chip capacitors and resistors) and the solder joints holding them to the board are not unduly stressed either during the fixing process or from subsequent mechanical and/or cyclic thermo-mechanical forces.

Typical situations requiring attention are:

- Where the board is bowed and is intended to slide to and fro along slots in a rack-mounted equipment. If the bow is significant, the act of pushing it along the slots is to straighten the board locally and to impart mechanical stress to solder joints that are near the slot. One such action may be enough to cause open circuits or at least to initiate micro cracks leading to subsequent failure in the field.
- Where the same board is not bowed, but has to be pushed into a stiff multi-pin plug and socket arrangement at the backplane. This action can cause the same type of problems as are mentioned in the preceding paragraph. Again, one push can be enough to bring failures.
- Where the board is mounted rigidly to a metal frame whose material has

a significantly different thermal coefficient of expansion compared with the board. In this case the temperature changes in regular daily use, i.e. switching the equipment on and off, can cause repetitive reverse stress cycles which degrade the soldered connections to surface mounted components. Eventually this can result in open circuit joints.

A further point in this context is that the stress excursions need not, in themselves, be large enough to bring failures, but may still propagate existing micro-cracks within chip ceramic capacitors which have already been initiated during the soldering process (cf. Chapter 17).

5.7 Product safety

The same considerations apply to ensuring that the design task has included all the relevant product safety aspects prior to subcontracting. These may range from the simple instruction to procure and affix relevant user safety labels, to a request for a full-scale hazard and risk analysis by the subcontractor.

In most instances, the latter would be expected of any competent design department, but if the subcontractor has a better knowledge of the type of product and its problems than the customer—and is willing (which may be doubtful)—it may best be done in that quarter.

An example of a typical analysis for a board assembly is given in Fig. 5.2 and for a completed television set in Fig. 5.3. Related data on the risk factors involved are given in Table 5.1. Training is required in the use of these documents.

Product liability points that require purchasing management attention at the customer–subcontractor interface are discussed in Chapter 18.

While the purchasing function may bear no direct responsibility for ensuring that all key design matters are addressed at the proper time, when the option of subcontracting arises it is important that the right questions are asked inside the company before orders are placed. In this sense purchasing managers should see themselves as 'long-stop' guardians of their companies' interests by raising the issues—if only because their own personal status in the industry might suffer if they are associated with a problem product!

Product Type No. 123-XYZ   **Analysis Date** 10.10.90.

Notes: Key to Rank No's, Table 5.1
Rank No's 2/5 indicate occurrence risk 2 and criticality risk 5. Worst combination is underlined.

RECOMMENDED ACTIONS
i) Customer advised to check fail-safe for chip capacitor short circuit
ii)
iii)
iv)

Analyst's signature D. Boswell

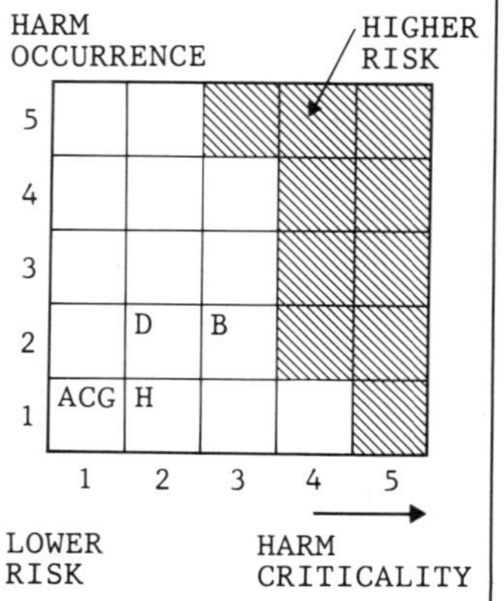

HAZARD LETTER	HAZARDS	Acceptance testing	Transport	Storage	Installation	Normal Operation	Misuse/Failure	Maintenance	Disposal	Other	CAUSES & COMMENTS
		PRODUCT LIFE PHASE RISK RANKINGS									
A	Electric shock	1/1	—	—	1/1	1/1	1/1	1/1	—	—	5 volts applied.
B	Fire	1/1	—	—	1/1	1/1	2/3	1/1	—	—	Is electrical design fail-safe for chip caps?
C	Explosion	1/1	—	—	1/1	1/1	1/1	1/1	—	—	Wound capacitors.
D	Toxic fumes	1/1	—	—	1/1	1/1	2/2	1/1	1/1	—	Plastic overheating above 200°C – fault conditions.
E	Radiation	—	—	—	—	—	—	—	—	—	
F	Harmful substances	—	—	—	—	—	—	—	—	—	
G	Cutting/Crushing	1/1	1/1	1/1	1/1	1/1	1/1	1/1	1/1	—	Negligible risk.
H	Surface Temperature	1/1	1/1	1/1	1/1	1/1	2/2	1/1	1/1	—	Refers to B and D.

HARM OCCURRENCE vs **HARM CRITICALITY**

HARM OCCURRENCE \ HARM CRITICALITY	1	2	3	4	5
5					
4					
3					
2		D	B		
1	ACG	H			

LOWER RISK → HIGHER RISK

Fig. 5.2 Safety profile: PCB assembly

		PRODUCT LIFE PHASE RISK RANKINGS								Product Type No. ABC-XYZ Analysis Date 17.02.92
HAZARD LETTER	HAZARDS	Acceptance testing	Transport	Storage	Installation	Normal Operation	Misuse/Failure	Maintenance	Disposal	Other	CAUSES & COMMENTS — Notes: Key to Rank No's, Table 5.1. Rank No's 2/5 indicate occurrence risk 2 and criticality risk 5. Worst combination is underlined.
A	Electric shock	2/5	—	—	2/5	1/5	3/5	2/5	—	—	With rear cover off.
B	Fire	1/2	1/2	1/5	1/2	1/5	2/5	1/2	1/5	—	Organic materials in cabinet and components.
C	Explosion	1/4	2/4	—	1/4	1/3	1/4	1/4	2/4	—	Damage to CRT.
D	Toxic fumes	1/2	—	—	1/2	1/3	1/3	1/3	3/3	—	Overheating or burning of organic materials.
E	Radiation	—	—	—	—	—	1/2	—	—	—	Not normally hazardous.
F	Harmful substances	—	—	—	—	—	—	—	1/3	—	Contamination of environment due to careless disposal.
G	Cutting/ Crushing	1/2	3/3	—	3/2	—	—	2/3	2/3	—	Handling into position.
H	Surface Temperature	1/2	—	—	1/2	1/3	1/3	1/3	—	—	With rear cover off.

RECOMMENDED ACTIONS

i) Hazard warnings for letters A, B, C, D, F, G, H.

ii) All organic materials to have Oxygen Index greater than 27.

iii) Fit cut-outs activated by smoke-detector, overload current, excess temperature.

iv) MD to receive & initial a copy of this Safety Profile.

Analyst's signature ...D. Boswell......

HARM OCCURRENCE						
5	▨	▨	▨	▨	I HIGHER RISK	
4	▨	▨	▨	▨	▨	
3				D, G	A	
2				E	C	B
1				F, H		
LOWER RISK	1	2	3	4	5	HARM CRITICALITY

Fig. 5.3 Safety profile: mains-operated TV receiver

Table 5.1 Harm occurrence and criticality ranking guidelines

\<Likelihood of harm occurring\>		\<Harm criticality\>	
Rank no.	Risk of occurrence	Rank no.	Severity of harm
1	Very low, e.g. 1 in 10 million chance	1	Negligible No injuries
2	Low, e.g. 1 in 1 million chance	2	Marginal No injuries, but minor property damage
3	Medium, e.g. 1 in 10 000 chance	3	Controlled Minor injuries not needing skilled medical attention Minor damage to property
4	High, e.g. 1 in 1000 chance	4	Critical Injuries need skilled medical treatment Serious damage to property
5	Very high, e.g. 1 in 100 chance	5	Catastrophic Reserved for most serious outcomes

6

Design checks before subcontracting hybrid circuits

6.1 Background comments

While thick and thin films represent mature technologies, unlike surface mount printed board assemblies, the layout of hybrid circuits is almost invariably carried out by the subcontractor. This is because the properties of the film components are highly dependent on the particular range of materials, process techniques and temperatures used. Each must be closely matched to the circuit requirement and specialist knowledge of the technology is essential.

For these reasons it is customary for the OEM to approach the hybrid circuit manufacturer well before electronic design commences so that best advantage can be taken of the benefits and the common technical and cost pitfalls can be avoided.

Manufacturers will want to make what appear to be substantial design and tooling charges, but most have tended to be flexible on the method of payment. Obviously, the size and time-scale of the ongoing production 'carrot' will be a vital factor in discussions.

Traditionally, manufacturers will expect to recover only about two-thirds of their real costs in making prototypes and the remainder over production runs. Apart from competitive market pressures, the reason for this apparent generosity is that they know that second sourcing by the OEM will involve a much higher level of extra costs than for alternative printed board structures. The hybrid house is therefore virtually assured of at least a substantial proportion of subsequent production—if the project succeeds.

The question of ownership of designs and tooling is similar to that discussed for surface mount printed board assemblies in Chapter 11.

6.2 Electronic circuit design constraints

- For thick film circuits, the designer should aim to keep the range of film

resistor values to a minimum—ideally so that only one ink resistivity is needed. Each additional resistivity requires an extra printing operation, widens the post-firing value tolerances achieved, and leads to more trimming and higher cost.
- Circuit design should be targeted so that the initial tolerances on thick film resistor value are suited to the technology, e.g. ±25 per cent or ±2 per cent—nothing in between. Using modern laser trimming methods, once the tolerances required are closer than those achieved *ex* firing, the latter being ±20–±30 per cent, depending on the number of resistivities used and the physical size of the resistors, the cost of adjustment to ±2 per cent by laser trimming is not very different from ±5 or ±10 per cent. Some suppliers are able to offer initial tolerances down to 0.1 per cent for certain parts of the ohmic value range. To justify such accuracy the ongoing stability and TCR must be controlled, and this may mean designing slightly larger resistor areas on the substrate.
- For thin film circuits, the ohmic value of film resistors should be kept to a minimum—ideally within the range 100–100 000 ohms. Normally only one resistivity is available on a given substrate and this may lie in the range 100–500 ohms per square. High-value thin film resistors need longer tracks and finer track widths, which add to costs because of the larger space occupancy on the substrate and reduced process yields.
- Lasers are also used to adjust resistor values while the circuit is operating. This enables dynamic trimming to precise functional performance, e.g. for filter frequency, current drain, gain or feedback levels. Besides improving the accuracy of these settings, this facility eliminates the need for resistor 'select on test' procedures.
- Film capacitors and crossovers are possible in both types of technology, but should be avoided where lowest cost or higher reliability is the main criterion.

6.3 Substrate layout and structural constraints

- The maximum safe size of a film circuit substrate is constrained by the brittle nature of the glass or ceramic used and the specified stress environment of the application. Cost is also a size-limiting factor when components are small and closely spaced and production yields are reduced in consequence.
- Within certain process and operating voltage limitations, thick film technology achieves miniaturization principally by making each resistor no larger than is absolutely necessary to achieve long-term parametric stability at the wattage dissipation it experiences while the circuit is functioning. This means that, in addition to ohmic values, initial tolerances and stability requirements, for the circuit designer to call up ¼ watt or ⅛ watt, as in the past, is no longer satisfactory and will add

unnecessary size and cost. For precision circuitry it will also be advisable to define a life calendar divided into periods at various temperatures and cyclic ambients so that stability can be designed to be adequate over the expected life of the product. These points are important because the directions of drift in ohmic value depend on the content and resistivity of the various inks used and on the aspect ratio of the resistor track. Low resistivity inks tend to reduce in value when diffusion from the termination material into the resistor area has a greater effect than oxidation of the resistor material itself. This is more likely to occur when the resistor has a low length-to-width ratio. Conversely, in higher-resistivity materials and with longer, finer tracks, oxidation is the dominant factor and the value increases with time. These matters are dealt with by the hybrid designer under the heading of 'end effects'. In selecting a supplier, it is worth asking to see the company's design rules to ascertain whether the above points are covered. Their existence will mark the difference between an amateur 'cut-and-try' outfit and the experienced professional team.

- If the circuit unavoidably requires a few crossovers, it is cheaper to specify soldered mechanical bridging links as added components. Alternatively, the terminations of added semiconductors and chip components may be used to provide the same functions.
- If resistor noise is a potential problem for a particular circuit, the thick film designer can minimize it by selecting suitable resistivities, track aspect ratios and trimming procedures.
- Subcontractors tend to limit the materials offered as hybrid circuit encapsulants to those in which they have developed expertise. Purchasers should ensure that what is offered is compatible with subsequent in-house processing, e.g. soldering, cleaning, testing and further encapsulation, and that it is suitable for the prescribed environmental specification.

All plastics eventually allow moisture penetration and hence of ionic contaminants. Circuits containing closely spaced fine tracks are more sensitive on this issue. However, contamination from external sources is often less of a risk than failure to clean such assemblies prior to encapsulation.

- Users intending subsequently to embed encapsulated or unencapsulated film circuits in a casting resin should evaluate inherent stress levels before designing in such arrangements. Cracked substrates occur when the structure is unbalanced.
- Where a hybrid circuit is purchased wholly or partly encapsulated so that the customer can later solder leads to designated conductor pad areas, great care is needed from the subcontractor, some of whose post-firing processes may reduce the solderability of such surfaces.

Note: When contamination arising from the curing of resin or other sources is unavoidable and all else fails, hard rubber erasers of the type previously used by office typists before the invention of correction tapes have been used to overcome the difficulty!

- The stresses arising from TCE mismatch between large ceramic chip capacitors and alumina substrates are different from those occurring in printed board assemblies. In this case, typical capacitors contain barium titanate dielectrics with TCEs higher than that of alumina. On printed boards, the capacitor TCEs are lower than, for example, those of typical epoxy–fibreglass materials. As with surface mounted printed board assemblies, the difference may create the need for use of conducting adhesives to replace solder in wide-temperature-range applications. When this is essential, the chip capacitors should be ordered without a solder coating on their termination areas so that good adherence is obtained and semi-conduction at interfaces is kept to a minimum.

7

Surface mounted components and assembly materials

Typical active and passive surface mounted component styles are shown in Figs. 7.1 and 7.2. These are discussed in detail in the author's earlier publications, listed in the Introduction to this book.

7.1 Solderability of components

The importance of good solderability has risen sharply for surface mounted components. For this reason it takes pride of place in this chapter.

As the solderability life of a plated lead is very dependent on the plating thickness, there is a need to identify this thickness before ordering for use in reflow-soldering situations. If the information is not given in the manufacturer's data sheet, the required minimum of 8 microns should be specified on the purchase order. However, manufacturers often specify 8 ± 3 microns, which means that the minimum is 5 microns, and this may not be acceptable, depending on the time since manufacture and the prior transit and planned future storage conditions.

In the absence of a guaranteed minimum solder thickness of 8 microns or of suitable international solderability tests for components to be assembled using reflow techniques, it is prudent, for example, to specify that 'component terminations shall be of an age and finish suitable for reflow soldering at high yield with RMA fluxes'. This should at least carry some weight in any post-mortem discussion on 'fitness for purpose'.

For the above reasons, many users specify for all surface mounted components that their date codes of manufacture shall be less than six months prior to the date of receipt by their goods inwards department. Given supply from a manufacturer of high-quality solderable termination surfaces with a solder coating 8 microns thick, this should allow a further six months' storage without major adverse effects on solder joint yield.

84 SUBCONTRACTING ELECTRONICS

(a) MELF diode   (b) SO diode (SOD123)   (c) SO transistor (SOT 23)

(d) SO thyristor (SOT 143)   (e) SO medium power transistor (SOT 223)

(f) SO integrated circuit   (g) SOJ integrated circuit

(h) Plastic leaded chip carrier integrated circuit (PLCC)

(i) Quadpack integrated circuit

(j) Miniature PLCC integrated circuit

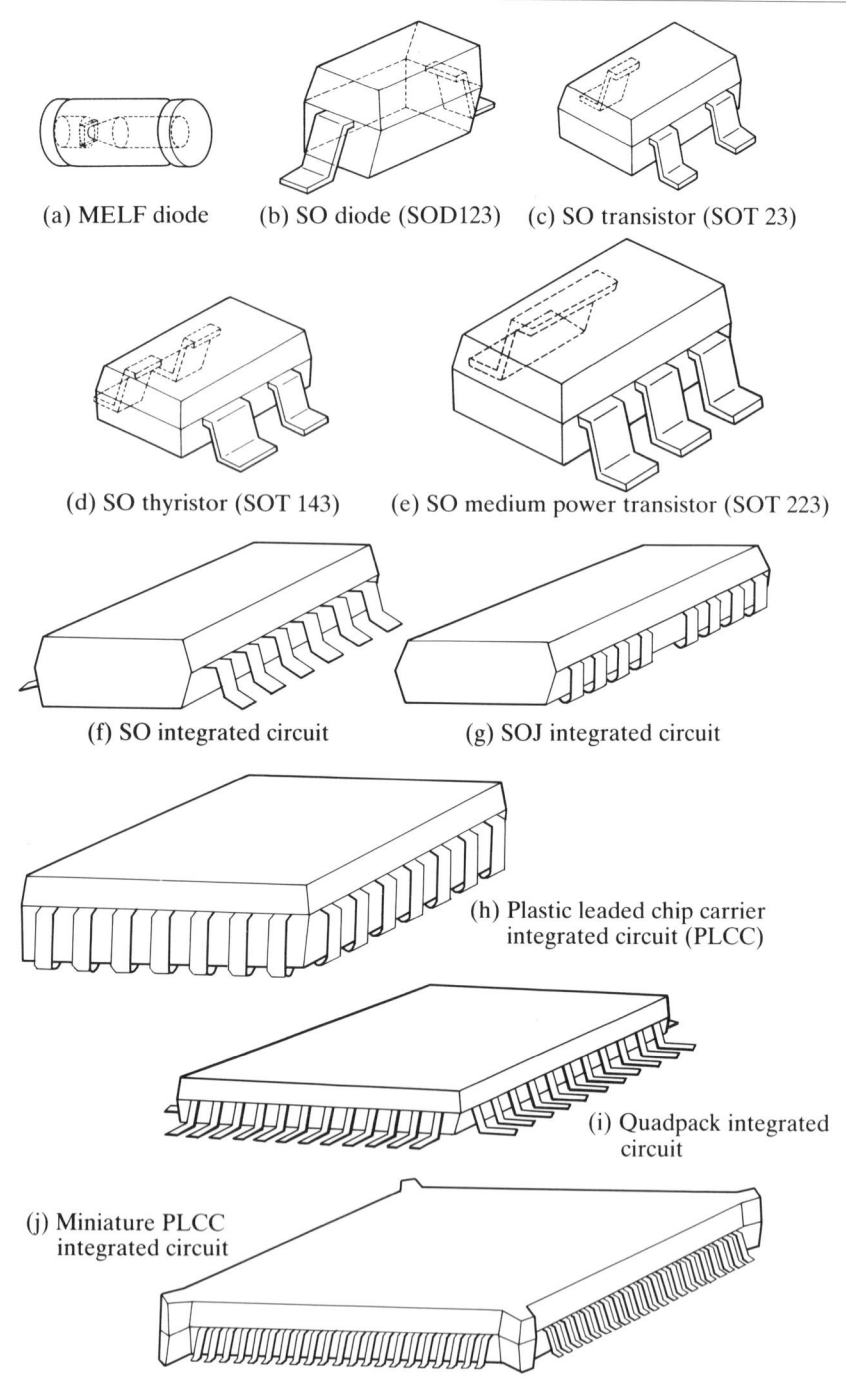

Fig. 7.1 Typical semiconductor surface mounted component package styles

SM COMPONENTS AND ASSEMBLY MATERIALS 85

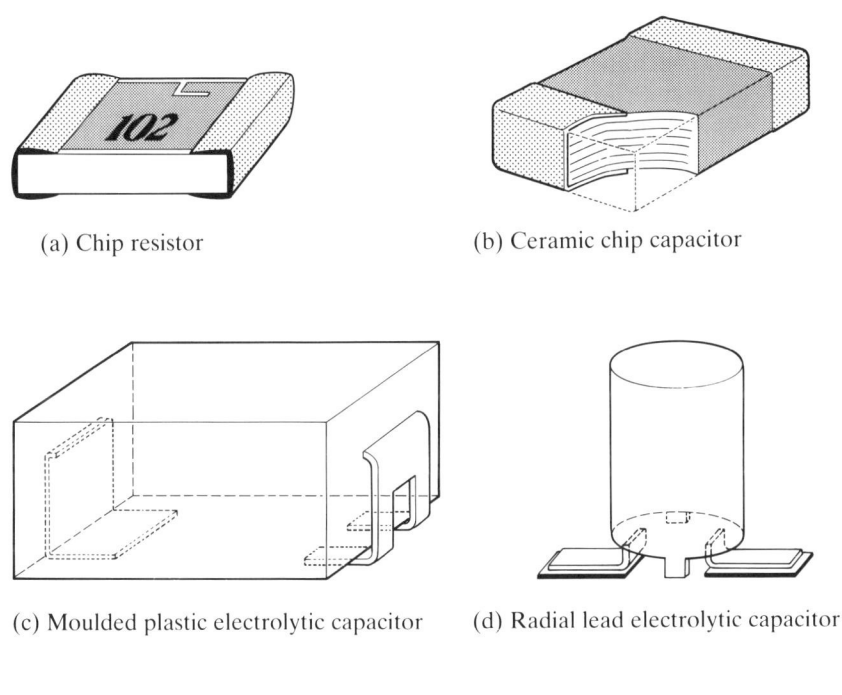

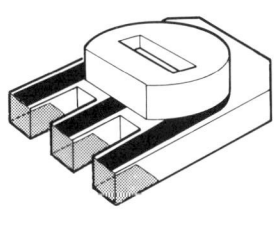

(a) Chip resistor

(b) Ceramic chip capacitor

(c) Moulded plastic electrolytic capacitor

(d) Radial lead electrolytic capacitor

(e) Chip resistor trimmer

Fig. 7.2   Typical passive surface mounted component styles

7.2   Specifying and purchasing surface mounted semiconductors

Figure 7.1 shows typical semiconductor surface mounted components. In addition to the normal requirements common to conventional leaded components, e.g. manufacturer's type number, etc., the following points require detailed consideration.

(A)   ELECTRICAL PARAMETERS

Most firms purchase semiconductors from their suppliers on the basis of 100 per cent testing prior to shipment. The electrical function of standard semiconductor devices is not usually tested at goods inwards unless it is necessary to check them over their maximum temperature range, e.g. for military or avionics applications.

However, although custom-designed application-specific integrated circuits (ASICs) are tested at goods inwards by larger OEMs—normally at room temperature—subcontract assemblers often do not possess the necessary equipment. In this case either the OEM passes tested devices to its subcontractor or the customer accepts the risk arising from a very simple in-circuit test.

Acceptance quality levels (AQLs) of less than 0.1 per cent should be applied on electrical parameter testing.

(B) MECHANICAL OUTLINE

In view of the lack of standardization, it is a wise precaution to specify all semiconductors against a manufacturer's drawing number as well as a national or international outline. The printed board layout should have been based on the former.

(C) DEVICE PACKAGE INTEGRITY

For larger integrated circuits in PLCC (plastic, leaded chip carrier), quad-pack and SO (small-outline) styles above 20-pin packages, precautions against the 'popcorn effect' are advisable, particularly where a new supplier's record in this respect is unknown (cf. Section 17.3).

In a few cases the potential for sudden cracking due to thermal shock during soldering, e.g. as a result of poor moulding, has been picked up by visual inspection at goods inwards. Packages housing large chips or having copper leads are reported as being more prone to this fault.

It can be very important to check whether the manufacturer's drawing offers a lead coplanarity specification that is suitable for the intended assembly process. Devices for reflow soldering should have the underside of their leads coplanar within 0.06 mm (0.0024 in). Where this is not stated by the manufacturer it should form part of the purchase specification, especially for larger packages with lead counts in excess of 40. For lead pitches at 0.5 mm and below, the checking of this parameter may require sophisticated laser profile scanning equipment with a sensitivity of a few microns.

7.3 Specifying and purchasing typical passive components

Figure 7.2 shows typical passive surface mounted components. In addition to the normal requirements common to conventional leaded components, e.g. parametric value, wattage, voltage, tolerance, the following points require detailed consideration.

(A) COMPONENT SIZE

Surface mount printed board pad sizes are defined in relation to specific chip component standard dimensions. However, capacitor chips of any given value can be supplied in a variety of different plan sizes, hence it is vital to specify not only the capacitor value, but also the plan size, e.g. 1206, 1812, for which the board layout has been designed.

When reflow-soldering is intended, it is also important to specify the thickness of chip required. There are two reasons for this. The first is that, for reflow, the amount of solder paste to be deposited on a given footprint pad size will not always be the same. Most visual inspection standards specify a solder meniscus height h which relates to the chip body height H, e.g. $h > 0.25H$. To meet this type of specification, a 1206 chip whose thickness is 2.0 mm will need more solder than a 1206 chip with a thickness of only 0.6 mm.

Secondly, there are different methods of manufacturing chip capacitors that result in different chip thickness for the same value capacitor having the same plan size dimensions. For example, a 470 pF 1206 size X7R capacitor from supplier A may be 50 per cent thicker than the same item from supplier B. Production yields will be adversely affected if these points are missed. As this affects the design of solder paste screens and stencils, the issue has to be made clear to printed board layout designers who should be responsible for screen/stencil design.

(B) PRIMARY PACKAGING

The primary packaging must be suited exactly to the component placement machine requirement, e.g. reel, tube or tray size.

Unless stated in the supplier's data sheet that all items are supplied in anti-static packaging, any specific anti-static requirements should be defined. Note that some such packaging materials may contain amines which can cause solderability to deteriorate.

Small quantities of components should also be specified on the order to be supplied either in part full reels or in cut lengths of tape off the reel, or in part-full tubes or trays. If de-taping or decanting is essential because no suitable hand feeder is available, this operation should occur at the same time as assembly so as to minimize exposure time to handling, factory dust, fumes, etc.

Packing of surface mounted components loose in polythene bags is now so widely regarded as being capable of causing solderability to deteriorate under reflow conditions that it should not be necessary to resort to specifying its prohibition on the order, although some users still do so as a precaution.

It is recommended in 7.4(A) that ceramic chip components supplied loose in polythene bags should be rejected without inspection. This is because the components rub against each other and fine particles of ceramic powder can become engrained in the solder. The polythene also rubs off on the rougher parts of the solder surface and is extremely difficult to remove. Both can reduce solderability. The same inter-component rubbing occurs when ceramic chips are supplied loose in plastic vials or cassettes.

7.4 Goods inwards inspection, storage and kitting of surface mounted components for high yield in assembly

(A) INITIAL CHECKS
- Reject all components supplied loose in any form, especially if they are loose in polythene bags or other packaging materials that may inhibit solderability.
- For items packaged in reeled tapes, on each reel check the component type number and/or the parametric value(s) in ohms/nanofarads, etc., against the order and ensure that the date codes of manufacture and reeling are within those specified on the order. Check the same points for components packaged in each of the tubes or trays or any other carrying media. If any of the these checks give cause for rejection, do not open the bag, tube, etc. Classify as 'Reject'.
- Count and/or weigh to determine the quantity supplied. For example, open a bag, remove the reel or select a tube or tray and weigh it against a previous specification weight range for a full one. Weighing can be in groups thereafter if the scales are accurate. Alternatively, there are machines that unreel and re-reel the whole tape, counting each pocket in the process. Note that many machines count the pockets rather than the components, thus assuming that there is a component present in each pocket. This may not always be so.

(B) DIMENSIONAL AND SURFACE CONDITION CHECKS
- Check the sample components for overall dimensions.
- In the case of PLCC and Quadpack styles, visually inspect for obvious lack of coplanarity of their leads and any other accidental deformation. If a rigorous coplanarity check is required, either the suspect device can be placed leads-down on a flat plate and feeler gauges used to estimate errors, or a laser scanning system capable of better than 10 micron resolution can be applied—if available.
- Pay particular attention to the outline distance over semiconductor integrated circuit leads and compare it with the maximum pad outline given on the printed board layout drawing. This is more important if high-pincount components are sourced from Far Eastern countries, owing to unsatisfactory standardization between manufacturers in that region.

At the same time, check for visual defects using a ×4 illuminated magnifier or a ×10 binocular microscope.
- On chip ceramic components, check the three samples for signs of leaching on the termination areas, cracks and general visual appearance, again using a ×4 illuminated magnifier or a ×10 binocular microscope.
- Chip capacitor edges should be checked for signs of exposure of the internal metallization layers arising from bad alignment, and for delamination which is usually indicated by body colour variation on the same chip. Resistors should be examined for damage to the (black) resistive track and its glaze covering.

(C) SOLDERABILITY
- If reflow soldering is intended, check all types of component for solderability if the date code shows that the components were made more than 6 months ago (12 months ago for wave-soldering). Use three samples freshly taken from the reel, tube or tray to avoid the risk that devices may have been accidentally contaminated during prior inspections.
- Note that passing the IEC 68–2–58 Part 2, Test Td by dipping SM component terminations in a solder bath and visually observing results will not ensure satisfactory results for reflow-soldering using RMA, no-clean or low residue fluxes. Where it is intended to stock components for several months, ideally the solderability check at goods inwards should include artificial ageing prior to the test, for example using dry heat as in IEC 68–2–2, Test B (155 °C for 4, 16 or 72 hours) or steam as in IEC 68–2–20, Test T (100 °C in 100% RH for 1 or 4 hours).

  Sectioning methods to verify solder thickness may be applied in critical cases. This involves the casting of components in a suitable resin followed by sawing—in some cases with a diamond disc—and very careful lapping techniques to avoid smearing the metals. Some capacitor manufacturers employ fracture sectioning to overcome the latter problem. The thickness of layers may then be determined using a metallurgical measuring microscope or an electron microscope.
- In large-volume production runs, it is customary to send samples from each delivery of suspect components down the production line to check for hidden defects or poor solderability. Many believe that this is the only valid determinant of fitness for purpose.

(D) ELECTRICAL CHECKS
- Some assemblers check diodes and transistors at goods inwards for parameters that are critical to the performance of their circuitry. The same can be said of complex integrated circuits where it is practicable to socket them into a test rig. On this point, note that badly designed test jigs or unsatisfactory operator handling procedures have often been the

cause of high post-soldering defect levels because of lead deformation or solderability deterioration prior to placement.
- For chip passive components supplied in reels, do not cut off the leader tape—which is needed to load the pick-and-place machine—but carefully peel back the transparent cover tape to remove the first three components from each reel. This should be done with conductive plastic (not metal) tweezers which should be applied to the body of the component rather than to the termination areas. Check that their parametric performance or value and tolerance match those stated on the manufacturer's data sheet or on the reel and that their visual appearance meets the relevant requirements. Some inspectors use needles to insert as electrical probes through the side of the tape to avoid removing passive components. This may enable parts at several points in the reel to be at least checked for parametric value.
- For close tolerance resistors and capacitors, e.g. ± 1 or 2 per cent, using the same method as in the preceding paragraph regarding chip passive components. As a minimum, check a sample of 10 for compliance within tolerance limits. Stricter visual standards are essential.

7.5 Adhesives and solder pastes

More companies have reported difficulties in handling adhesives and solder pastes than for any other subject except solderability. There are three main issues: shelf life, pot life and process conditions. The first two are within the province of the purchasing and stores functions.

(A) THE GOLDEN RULES FOR PURCHASING AND STORAGE
- Specify the right solder content and viscosity for the process. Specifications will be different for each of the following production methods of deposition: screen printing, stencil printing, syringe dispensing, pin transfer printing, fine line printing.
- Specify a container size that ensures complete usage within 48 hours.
- Specify that the date code of manufacture shall appear on each individual container and that details of how to interpret the date code are supplied with each delivery.
- Rotate stock on a strict 'date code of manufacture' basis. Do not use simple FIFO methods.
- Store in manufacturer's recommended conditions. Cooling is acceptable; freezing is not. Some solder paste syringes benefit from periodic physical rotation to avoid the settling out of metal particles.

(B) THE GOLDEN RULES FOR HIGH-YIELD PROCESSING
- Ensure that materials reach the deposition equipment at the correct ambient temperature—expecially if they have been stored in a refrigerator.
- Do not use materials whose shelf life has expired.
- Do not touch adhesive dispensing syringe nozzles with fingers. This can initiate curing.
- Deposit adhesives and solder pastes under controlled temperature conditions. Avoid direct sunlight.
- Do not put solder paste from the screen/stencil back with fresh paste in the post unless it is certain that all of the mixture will be used within 24 hours without further additions to the pot.
- Do not attempt to dilute a pot of solder paste that is too viscous to deposit properly. Return it to stores as reject material.
- Avoid deposition in dusty conditions.

8

Printed boards

8.1 Specifying and purchasing surface mount printed boards: a checklist

[*Note*: For the reader's convenience, an updated version of text from the author's earlier books is reproduced here. Reference to these books is made in the Introduction.]

As part of the purchasing specification for a printed board, a list comparable to that given below should be prepared and its data agreed with the supplier. This is not intended as a complete quality guideline, but it indicates most of the features specific to surface mounting and mixed-technology requirements and can form a basis for negotiation.

Many of the points covered are best dealt with in a general specification to be used for all surface mount printed board purchases. Other requirements will be specific to individual products.

(A) SUPPLIER'S 'FRONT END' SYSTEM
More printed board manufacturers are now equipping themselves with what they call a 'front end' system. This enables them to scan a customer's photomaster, create a tape or disk, and use a computer to make automatic correction to their factory artwork (photo-tools) to compensate for etching factors, particularly for fine track widths and gaps. Where a company offers this facility, it is essential not to make allowance for etching, but to design on the CAD system and to specify the actual finished track widths required.

The purchasing function should ensure that the printed board designer is aware of the intended supplier's capability in this respect before he starts a layout on the CAD facility so that, if no front end system is available, suitable allowances are made on the designed track widths and gaps. This is important for track widths at 0.2 mm and below. For example, where a photomaster shows a long, thin track that is 0.125 mm (5 mil) wide, the finished product may be as narrow as 0.075 mm (3.5 mil) and have a DC resistance 30 per cent higher than was intended.

(B) PHOTOMASTERS, FLOPPY DISKS AND DIMENSIONED DRAWINGS
In most cases it is no longer satisfactory merely to send a magnified or 'at size' photomaster (or the equivalent on a floppy disk) to a printed board manufacturer. First, this will not necessarily be used for board production, and second, unless the photomaster or disk/tape supplied is accompanied by dimensioned drawings indicating the relative accuracy requirements, the supplier will have insufficient data on the necessary precision for surface mounting. Refer also to (L) below. This stricture applies equally to the accuracy of step-and-repeat arrays.

Nor is it wise to supply data solely on disk. Dimensioned photomasters or printouts generated from the disk should accompany it, because then it is comparatively easy for the board manufacturer to check them for hidden errors and against the real requirement.

It is vitally important for the quality department to understand and ensure effective control of the different dimensioning and tolerancing protocols for boards populated on pick-and-place machines using dowel pin location datums versus optical correction mark datums. Refer to (M) below.

The printed board layout designer should know the alignment capability options of the placement machine he is designing for. The quality manager should ensure that the drawings sent to the board manufacturer reflect reality and are dimensioned with reference to the correct datums.

(C) SPECIFYING THE BASE MATERIAL
Specify the base material and its thickness, e.g. FR4, copper-clad epoxy fibreglass (e.g. BS 4584 Pt 2, Close tolerance). Tolerances on multilayer boards are wider:

0.8 mm thick	±	0.10 mm
1.6		0.15
2.4		0.18
3.2		0.20

Note that 'FR' stands for the degree of fire resistance of the material. There are a number of quality and cost options available within the FR4 range that reflect factors other than fire resistance, depending on user requirements. If the purchasing specification merely calls up 'FR4', the supplier is likely to offer the cheapest version of that range that he has in stock, and this may not necessarily have sufficient dimensional stability for use with surface mounting processes.

(D) INITIAL AND FINAL COPPER THICKNESS

For long, fine tracks, it may be necessary to define the maximum ohmic resistance between two points on the board because that can no longer be assumed to be zero in all cases. This highlights the importance of specifying the final copper thickness as in (C) above and of suitably tolerancing the width of fine tracks. Discuss initial copper thickness, e.g. 5, 10, 17.5 (½ oz), 35 (1 oz) microns, but specify the final thickness, e.g. 15, 25, 30 microns, especially if fine tracks are used.

(E) COPPER TRACK ACCURACY

Tolerances on track widths and gaps should be specified, especially for those below 0.2 mm (8 mil).

(F) SOLDER RESIST MATERIAL

Production yield and reliability are affected by the choice of resist material and its thickness. There is also the question of ensuring that the selected resist is compatible with both the soldering process and other materials in contact with it, for example adhesives, cleaning fluids, gases used in soldering.

Today the main resist options for good results include wet film photo-imageable direct on copper or copper oxide, and dry film photo-imageable also direct on copper. For UHF and VHF work, nickel plating is often preferred.

The conventional low-cost option applied in the past to through-hole assemblies is to have a screen-printed resist over the entire solder-coated board in which apertures appear where solder joints and test points are needed. This is unsuitable for most surface mount assemblies because the adhesion and alignment accuracy are both poor and the fact that there is solder underneath the resist means that control over the contours of small solder joints is also poor. Surface tension can drag solder away from the joints to form thickened tracks beneath the resist layer. This form of resist softens during the soldering process, and the final effect is a wrinkled surface which is both cosmetically and technically undesirable.

Also to be specified are the necessary resist alignment accuracies in relation to solder pads and the clearances around fiducial marks and plated-through holes. Resist colour may be important in some applications.

(G) MULTILAYERING DATA

Specify the acceptable thickness range for each insulating layer and the co-alignment of specified copper pads on all separate layers. Typically, the latter accuracy would need to be within 0.06 mm (2.2 mil) and all tracks/

pads on individual layers within 0.04–0.1 mm (1.6–4.0 mil) of specified positions relative to datum, depending on board size.

(H) SOLDERABLE SURFACES

With the ozone layer depletion problem forcing progressive moves towards the use of low-activity fluxes, the importance of achieving greatly improved board solderability cannot be over-emphasized.

As with components, there are no 'magic wand' solutions. The best results come from good production practice, informed purchasing procedures and disciplined treatment of boards from date of manufacture through to the final soldering operation. The buyer has a key role in this sequence.

Specify the metal or solder coating requirement. Today the conducting surface material finish options include bare copper, bare copper with an oxide formation reducing coating (e.g. Entec), a fluxed lacquer, nickel/gold plating, hot gas or hot liquid levelled 60/40 tin/lead, tin/lead and plated tin/lead. The respective uses of each are as follows.

1 Bare copper and oxide-delaying coatings are used mainly where a solder paste is deposited prior to reflow-soldering. Boards purchased with unprotected bare copper should not be stored for more than one or two weeks as oxide formation inhibits effective wetting. For the same reason, boards with Entec coating should not be stored for longer than three to four weeks.

2 Hot gas-levelled surfaces are the most widely used finishes, though the degree of coating flatness may not be adequate for reflow-soldering fine pitch leads. Control over the final coating thickness is not one of its advantages and the range offered may well be 5–50 microns. For boards needing storage for more than a few weeks, the 5 micron thickness may be insufficient to ensure good solderability under reflow conditions. In this case the minimum should be specified as 10 microns.

When wave soldering of glued surface mount components is intended, it is important to specify the maximum bump height allowed on pads receiving chip components. This factor assumes greater prominence when thicker resist layers are used—for example 100 micron dry film resist. The problem lies in a so-called 'seesaw effect', in which the edge of the resist becomes a fulcrum and the chip body is tilted by a local solder bump so that the adhesive is out of contact with its under-surface. Usually a tolerance of 0.04 mm on the maximum local protrusion height is appropriate.

3 Hot liquid levelling is achieved by passing the board very rapidly through a bath of liquid whose temperature is above the melting point of the solder used. It is applied where good flatness of the solder coating is required and the boards will be used within a few days after the levelling process

has occurred. This is advisable because the thickness of the solder is limited to about 3 microns.
4 Brushed tin/lead surfaces are marginally flatter than hot gas-levelled surfaces. However, the exposed surface area is greatly increased by the act of brushing, with a resulting increase in the amount of oxide present. For this reason, it is advisable that soldering takes place shortly after brushing—typically, the interim storage time should be less than a week. Care is needed to avoid distributing tin/lead slivers during brushing.
5 Tin/lead plating is a relatively slow process compared with solder dipping and is generally less favoured because of the consequent higher cost.
6 A nickel plating topped with a 1–2 micron thick gold flash is now used where the pad surface flatness requirement is critical, e.g. for high pin-count packages with lead spacings below 0.8 mm. In this connection, the risk of embrittlement of the joint arising from its progressive diffusion into the solder can be high, and calculation should be made of the effect on the critical trapped volume of solder lying beneath the termination of chip ceramic components. The percentage of gold in this region should not exceed 4 per cent as measured after soldering.

(I) NON-SOLDERED SURFACES

Specify the thickness and location of any special platings for edge connectors or contact pads that are not required to be soldered, e.g. gold, carbon, nickel.

Note that, for surfaces that are intended to receive semiconductor bond wires as in 'chip-on-board' structures and those needed for regular edge connector insertion/extraction, the hardness of the plating, e.g. gold, is of vital importance and must be specified.

Note also that it is unwise to use resist materials as crossover insulation for carbon materials—the point being that it is best to leave battery manufacture to those who understand it.

(J) STEP-AND-REPEAT REQUIREMENTS

Professional board manufacturers welcome pre-layout discussion on step-and-repeat requirements, including routing slot widths and break-out lug widths. The latter should be specified, as excessive width can cause damage to nearby surface mount joints during break-out. Their accuracy, related to the copper pattern and/or fiducial marks, should also be part of the specification. Refer to (M) below for details of dimensioning requirements.

(K) PANEL AND INDIVIDUAL BOARD SIZES

The panel size and accuracy tolerances can be very important where there is mechanized pass-through of boards on an automatic pick-and-place machine. Again, refer to (M) below.

(L) JIGGING HOLES

Noting the point made in (K) above, these should be specified as to their positions, sizes and positional tolerances related to the rest of the copper pattern and, where appropriate, to fiducial marks. The dimensions would be defined for both panel (process alignment) and individual boards (alignment for electrical test).

Jigging holes must not be plated-through.

Buyers should be aware that surface mount component assembly demands kinematic design of location systems in order to get the best accuracy. This means that the tradition of specifying two—and sometimes three—round dowel holes is inadequate, as any slight dimensional inaccuracy arising from board manufacture can lead operators to force misaligned holes on to dowels, thus causing unnecessary printing or component placement errors. The correct design procedure is to specify one hole and one slot so that the misalignment mentioned will not cause serious placement errors.

For accuracy and repeatability reasons, it is recommended that preference should be given to suppliers who have a single machine that will drill and rout the location holes and slots without demounting the board and who are prepared to do this prior to etching copper so that they can be used for alignment. This means that the location holes require masking during the plating-through operation, which adds marginally to cost.

Some assemblers prefer the supplier not to utilize the user's jigging holes during board manufacture because of the risk of damage. Others allow this on the grounds that if anything is in error, hopefully the supplier will find the problem before shipment!

(M) FIDUCIAL (OPTICAL ALIGNMENT) MARKS

When the boards require optical alignment for screen-printing solder paste or auto-placement of components, the surface finish and positional accuracy of fiducial marks in relation to each other and to other key points must be defined by the layout designer. The achievable accuracy will depend on board size, but typically all related locations should be within 0.04–0.12 of their correct position.

Some placement machine optical recognition systems prefer solder coating on the marks for optimum contrast, while others react better to bare copper. The fact that not all machines recognize all mark patterns means that the board layout designer should be told the exact make and type of pick-and-place machine to be used. In turn, this implies that the purchasing department should select any intended subcontractor before layout commences—and there are many other reasons why this action is important.

Board designers used to dimensioning copper positions in relation to dowel hole locations often forget that, once optical recognition is applicable, the location of all copper areas, location holes and break-out slots should be dimensioned with respect to the fiducial marks, not to the dowel holes. The location holes may still be needed to stop the board moving around during auto-placement, but their positions must now also be defined in relation to the fiducial marks. Failure to recognize this situation can result in forced acceptance of boards that meet their accuracy specification, but on which the positional errors are double what they need to be.

On placement machines that scan two or three fiducial marks and effect placement on the basis of halving errors, for best placement accuracy, component pad positions should always be dimensioned and defined in relation to their nearest fiducial mark. In this case there is more than one datum.

Table 8.1 indicates what should be expected from a competent surface mount printed board manufacturer using ½ oz copper on FR4 material. The

Table 8.1 Accuracy of printed board copper pattern locations

Direct distance of copper pad centre line from datum (cm)	Radial error of copper pad centre from correct position ref dowel hole datum (mm)	Radial error of copper pad centre from correct position ref fiducial mark datum (mm)
5	0.09	0.04
10	0.10	0.05
15	0.12	0.06
20	0.14	0.07
25	0.17	0.09
30	0.20	0.12

figures do not necessarily correspond to the true requirements, which can be more stringent, particularly in the case of alternative substrate materials and/or very high-frequency circuits.

Examples
1 For a board not intended for optical correction and having a pad centre line spaced 15 cm and 25 cm from datum lines, as indicated in Table 8.1, the zonal error envelope diameter for that pad centre would be 0.365 mm, derived from

$2 \times \sqrt{0.122^2 + 0.172^2}$ mm = 0.365 mm.

2 For a board designed for optical correction and having a pad centre line spaced 10 cm and 15 cm from datum lines, as indicated in Table 8.1, the zonal error envelope diameter would be 0.156 mm, derived from

$2 \times \sqrt{0.052^2 + 0.062^2}$ mm = 0.156 mm,

thus approximately halving the error compared with dowel hole methods.

From the above data and from experience, it is suggested that:

- The pick-and-place machine will need to be programmed to take account of the initial errors. However, once programmed, what will be the likely variations from the original settings as between printed board batches? From the same supplier, the variations could be between 30 and 60 per cent, and from different suppliers, between 50 and 80 per cent of the figures given in Table 8.1, depending on the skill and quality control ability of the board manufacturer. This is of less consequence when optical correction is applied, but can otherwise result in continual juggling with the placement programme each time a new batch of bare boards is used.
- The dimensional methods employed should be aimed at optimizing repeatability rather than achieving absolute accuracy in relating to datums.
- Where optical correction is used, the accuracies of copper-to-copper relationships are more important than copper-to-dowel hole precision.
- As previously indicated, incorrect relational dimensioning can result in the supplier shipping product conforming to drawing, yet on which many of the above errors are doubled.
- It should not be forgotten that for FR4 material, 300 mm (12 in.) long circuits made by the printed board manufacturer at a temperature of, say, 15 °C, will be 0.05 mm (2 mil) longer if the user's assembly shop ambient temperature is 25 °C.

(N) VIAS AND PLATED-THROUGH HOLES

All hole and land diameters should be specified after plating, e.g. 0.5 mm (20 mil) via diameter with 0.025 mm (1 mil) minimum plating thickness, pad diameter 1.1 mm (44 mil). Tolerances should be agreed with the supplier, and it is also vital to specify whether any vias require tenting with resist.

(O) BOARD DISTORTION

Surface mount assembly processes and the reliability of resulting assemblies require more stringent control of board flatness than through-hole equivalents. There is disagreement in the trade about the achievable flatness versus the needs of surface mount process equipment; the arguments are stated in Section 8.2 below.

A common compromise is to specify a maximum pre-assembly distortion for both panels and individual boards of 0.4 per cent. A typical supplier response is to say that, while they cannot guarantee it, most of their boards will be within this limit.

While it is always possible to flatten boards by heating them when there are weights holding them down, internal stresses are relieved by the soldering operation and they then resume their distorted condition. Excessive bow can also cause SM assembly reliability problems during subsequent handling; see Section 17.1.

Purchasing managers should encourage board layout designers to study and apply the available guidelines for minimizing bow and twist. These are identified in the Bibliography (page 259).

(P) TRACK DEFECTS AND REPAIR METHODS

The maximum allowable track defect characteristics assume greater importance for boards carrying fine lines. For track widths at 0.25 mm or less, not more than 25 per cent of specified track width should be missing for more than 1.0 mm length. For track widths above 0.25 mm, not more than 30 per cent should be missing for more than 2.0 mm length.

Again, for fine lines, to protect reliability the allowable repair techniques must be different from those seen in conventional boards. These and any limitations on the permitted number of repairs per panel/circuit should be specified. The latter applies particularly for multilayer boards.

(Q) IDENTIFICATION MARKING

The normal method of applying identification ('ident') marking is via a screen-printing process which is often less accurate than is required for surface mount boards. The essential need is to avoid its encroachment on to any pad on which a solder joint is to be made, and for this reason the general advice is to avoid its use where surface mount components are to be assembled. The printed white boxes, which were essential in identifying the particular through-holes relevant to each wire-ended component, are a hangover from the past. They are no longer needed for surface-mount items. Where ident is essential, the material, its thickness, the method of application and all alignment requirements should be defined.

It is also worth while ensuring that the design department has made sure that the selected identification material will not adversely affect, or be affected by, any of the intended assembly and cleaning processes.

(R) BARE BOARD TESTING
Specify the calibre of bare board electrical testing required.

A 100 per cent electrical test by either the supplier or the user is considered mandatory for surface mount and mixed-technology bare boards. While most board manufacturers offer a 'bare board test', often they mean only a partial test confirming the tracks on each outer surface without the effectiveness of an interlayer/through-board connection check. The latter requires at least a shorting plate, and in some cases, where flatness is a problem, a 'clamshell' fixture—which is expensive. For large boards and those containing many vias (e.g. > 100), a test that exercises the complete board structure in accordance with the 'net list' is strongly recommended because of the potential cost savings at the end of the production line.

In this context, the fixture used for assembled board testing can sometimes be used for checking bare boards at goods inwards—if additions to the software and a few extra test points are designed in for the purpose.

(S) DEFECTIVE CIRCUITS IN ARRAYS
For panels with step-and-repeat arrays, specify the number of defective individual circuits allowed and the method used to identify them. Some cheaper pick-and-place machines do not offer a facility for the operator instantly to instruct that specific circuits in a step-and-repeat array must not be populated. This can lead to component wastage if the array is not 100 per cent good. On the other hand, the board manufacturer's production yield will be lower if the 100 per cent requirement is specified, with consequent cost implications which can be severe when there are many layers and a high percentage of fine tracks.

(T) TEST COUPONS
It may be helpful to agree with the supplier (and to specify) the format and location of any test coupons to be used by the supplier as a process and/or alignment monitor. These can then be incorporated into the CAD operation if the supplier prefers. Normally they are sited on the outer frame and broken away for test as part of the printed board manufacturer's final inspection procedure.

(U) INTENDED ASSEMBLY PROCESS PARAMETERS
Within the purchase specification, it is prudent to give the printed board manufacturer details of the intended assembly and cleaning processes,

including times, temperatures and surrounding materials. For life-support and long-life applications, a surface insulation resistance test (SIR) or ion count method should be specified to control residual contamination levels.

(V) ISSUE NUMBER AND DATE CODING METHOD
Specify the issue number and date coding method and their locations on the board. Preferably these should be etched in copper.

(W) BARE BOARD TRANSIT PACKING
Specify the packing requirements, e.g. non-hygroscopic; low free ion content (notably sodium, chlorine, sulphur); lint-free materials between boards; stacks to be packed in polypropylene sheet. Specify the maximum number of boards in one stack.

Some good suppliers offer heat-sealed dry-pack containers for multilayer boards.

Copper surfaces should not be allowed in direct contact with the polypropylene or with polythene, 'shrink wrap' or 'cling film'; the latter two in particular are not classed as low-ion-content materials, and in the presence of moisture they will deteriorate solderability.

The maximum number of boards packaged in one stack should also be specified. Badly bowed boards can be hidden at the centre of a large stack and may not be discovered until the batch is on the production line.

8.2 Bow and twist in large printed boards

The rising percentage of printed board area covered by copper resulting from the use of miniaturized surface mount assemblies with high-density track and pad content, and the wider use of multilayers, have increased both the occurrence and the magnitude of bow and twist in bare boards.

Lack of flatness has brought greater difficulty in carrying out important surface mount board assembly processes—especially in depositing solder paste and adhesive mounds of even height across large board areas, in component placement, and in some instances in soldering.

The international specifications accepted by printed board manufacturers require that the combined effects of bow and twist shall not exceed 1 per cent of maximum board dimensions. For bow the latter is taken as the length of the longest side, and for twist, that of the diagonal.

These limits were set years ago for conventional through-hole board assembly methods and are not suitable for surface mount technology. The surface mount equipment makers say they need a limit of 0.25 per cent, but the board makers say they cannot even guarantee 0.5 per cent and often blame poor layout design for the problem. Sometimes they are right.

Board flatness is affected by the layout pattern of the copper, by the layer thicknesses, by the quality of board material and by the methods used to make the bare board. The methods used for solder coating and for transport packing can also impair flatness.

To maintain the best flatness, among other precautions the designer must ensure good copper balance; preferably, all large earth and power planes should be cross-hatched rather than solid. The board maker must use the correct combinations of directionality of the embedded fibre. Unfortunately, these precautions cannot guarantee perfect flatness, but will give significant improvement. They may also result in slightly increased bare board prices, but the extra is likely to be more than offset by reduced assembly costs.

Engineers using CAD equipment for design of large surface mount boards should receive training in the art of minimizing the risk of bow and twist. Many engineers believe that the habit of specifying boards at 1.6 mm thickness for densely populated assemblies larger than 15 cm square needs to be reconsidered.

A draft international standard document giving evidence on how to optimize board flatness is referred to in the Bibliography (page 259).

8.3 Goods inwards inspection and storage of SM printed boards for high yield in assembly

The increased board complexity and component count inherent in applying surface mount technology have made it economically sound for assemblers to be more rigorous in their inspection and testing at goods inwards. The penalty for failing to carry out inspection and for finding defective boards only after assembly is now classed as an unacceptable financial risk by many who handle large-area boards in production.

Some aspects are more important during prototyping and pilot production phases, when there may still be design and specification errors or misunderstandings turning up and which require correction. In these circumstances there will need to be detailed dimensional as well as electrical checks, especially when optical correction is applied in assembly.

(A) CONFORMANCE WITH THE 'NET LIST'
Delivery of a batch of boards having a new circuit design should first be checked by the design department for conformance with the photomasters and/or other known design requirements.

(B) DIMENSIONAL VISUAL CHECKS
1 The distance between dowel holes can be checked by a jig simulating the supporting platen of the printer and/or the pick-and-place machine or, as for the distance between fiducial marks, with a travelling microscope.

2 The latter method would also be applied to the checking of key copper pad locations on the board in relation to the dowel holes or fiducial marks. The positions on the board and lead spacings of fine pitch integrated circuit footprints and those of small electrical test probe pads are all typical of those needing examination. Critical track widths and gaps should be inspected with the same microscope.
3 Panels having a step-and-repeat pattern should be checked for dimensional accuracy and compliance with rules governing the number of defective boards allowed within one panel—if any.

(C) VISUAL CHECKS

Sample visual checks on plating-through quality for vias should be carried out using a special large-depth-of-focus microscope or projector.

Vias used on SM boards are generally much smaller in diameter than those used conventionally and can give process control problems for some board manufacturers.

1 On the same samples, visually check the appearance of areas having fine lines and clearances. If in doubt, use a travelling microscope to measure their widths at both whole and any partially thinned regions.
2 Check the stack of panels/boards supplied for excessive gaps indicating excessive bow and twist likely to affect solder paste printing and pick-and-place machine operations. Measure likely defectives, e.g. by holding the board vertically against a right-angle surface table and inspecting with a dial gauge or with feeler gauges. Be careful not to touch any solderable surface with fingers or any greasy article.

(D) SOLDERABILITY

Check the solderability of sample stock made more than three months ago. To ensure that the manufacturer is not shipping old stock, a date code marked in copper should be a requirement in the purchase specification.

(E) CLEANLINESS

Where appropriate, check samples for contamination. Panel and board types most prone to contamination problems are those having:

- Many step-and-repeat circuits separated by milled slots
- Many vias
- Close fine tracks and gaps in the copper layout, especially those sited beneath large PLCCs or quadpacks
- High voltage requirements
- Need to operate in damp environments
- Usage in long-life, inaccessible equipment

8.4 Long-term storage and baking of printed boards
To avoid risk of blowholes and delamination due to outgassing during surface mount soldering operations, double-sided and multilayer boards that have been, or are likely to be, held in stock for more than one month should be dried out and sealed in polypropylene bags with sachets of dessicant, or baked just before use. This is particularly important for boards having large unbroken areas of earth plane or power plane.

For the drying or baking operation, a lower temperature for a longer time (e.g. 80 °C for 48 hours) is better than a higher one over a shorter period (e.g. 100 °C for 4 hours). The longer period is preferred to reduce the risk of bow, minimize the formation of oxide on the solder, and allow more time for the moisture to escape sideways in the plane of the board.

9

Components supplied by the customer

9.1 Free-issue scenarios

MONITORING USAGE

For the subcontractor, at first sight the main benefit of free-issue components from the customer is the avoidance of potentially severe cash flow problems. The materials are, after all, by far the largest element in the total board assembly cost. In practice, however, there are many offsetting disadvantages.

If the customer is experienced and professional in supplying on a free-issue basis, some of the advantages will remain. If not, then production planning by the subcontractor will be adversely affected and costs increased.

A subcontractor will be reluctant to set up machines for automatic SM assembly until all parts are available to complete the production batch run. This makes good sense, because later additions by hand are costly, less reliable and anathema to the production line system. (Refer to the example quote for production quantity in Chapter 13.) This means that customers wishing to delay deliveries for their own commercial inventory reasons can create major cost increases and hiccups in the subcontractor's scheduling and output programmes, merely by withholding a very few items.

Apart from noting that the subcontractor's stores are organized to maintain effective separation between projects and between individual customers' free-issue materials and general stock, the provisions for periodic inventory reporting should be noted. In most situations, a regular monthly report by the subcontractor is sufficient, with contract provision for prompt return (or sale to the subcontractor) of obsolete or surplus items.

An important factor is that the methods of presentation of components to pick-and-place machines make it impractical to issue kits to the assembly line having exactly the required number of components of each type to complete the batch run. Furthermore, any replacements needed as a result of electrical testing and rework may not be readily culled from tape reels fitted to automatic machines.

Inevitably, there are yield losses arising during both set-up and process optimization activities. It does not take many seconds for a fast pick-and-place machine to distribute quantities of components over the platen or on to the shop-floor! These may well be contaminated and should never be used—not even for rework. If they are unmarked ceramic chip capacitors, it is a fair bet that a percentage of those retrieved will have been lying around for several days in nooks and crannies as a result of previous malfunctions, and, although being of similar size and shape, they may have different parametric values.

In the 'quality costs' context, effective separate accounting of extra components used for set-up or rework is a very important issue. For this reason, well-managed subcontract operations devise paperwork or computerized systems which ensure that any extra parts needed because of poor yield are formally requisitioned and signed for against specific job numbers.

All part-used free-issue reels/tubes/trays from production other than rework should be returned to store for safekeeping and inventory control at the end of each batch run. It is not generally satisfactory to allow excess quantities left over after auto-placement batch runs to be kept on the shop-floor against a 'rainy day'. An auditor would look for a totally separate supply channel from stores for all components needed for rework operations.

Whatever the official company policy may be, there are, in these circumstances, strong temptations for shop-floor supervision and operators under output pressures to adopt 'Robin Hood' type tactics, i.e. to rob one customer's stock in order to maintain shipments on another's product. Inevitably, this is done in the hope that promised deliveries will materialize in time to restore the loss.

Apart from strict management discipline, one effective deterrent is for the customer to reserve the right to audit the position of free-issue items and purchased materials without prior warning. If the customer has paid in advance for items procured by the subcontractor specifically for the project, these could be made the subject of a similar audit right, but if they have not yet been invoiced and paid for, then, legally, they are still the subcontractor's property.

There is also the question of the method used by the subcontractor to minimize machine reloading times. Some keep a special group of component feeders which contain parametric values that are common to most circuits, e.g. 1000 ohm chip resistors, and 1000 pF and 100 nF chip ceramic multilayer capacitors. Normally these are supplied continually from store and their feeder positions on the machine are never altered. If the supply of a complete kit of free-issue components bespoke to one customer needs more spare feeders than are available outside the special group, the latter is dismantled.

Purchasing and subsequent kitting quantity procedures must allow for the above facts of life on the assembly line—but the targeted excess issues should always be quantified at the design or production engineering assessment stages, rather than be left for purchasing to decide.

INSPECTION OF FREE-ISSUE ITEMS

Whatever arrangements are made to cope with the problems discussed above, there is an absolute need to ensure that incoming materials are subjected to goods inwards inspection by an informed and properly equipped workforce who know what to look for—particularly on components and printed boards intended for surface mount assembly.

If this action is not performed by the customer before the goods are trans-shipped to the subcontractor, then the latter should be asked—and paid—to do the job. The subcontractor should be given sufficient time to carry out the inspection before putting the materials down the assembly line, with a negotiated price increase permitted if lower-than-budgeted yield (or worse) occurs because it is necessary to meet original delivery times which didn't allow for the inspection.

The purchasing contract should be crystal clear as to both the procedures and the respective responsibilities in the event that the customer provides free-issue items that are not 'fit for purpose' or do not allow the subcontractor to make assemblies within a reasonable percentage of costed yield. Refer to the typical quotations for production quantities given in Chapter 13.

The same approach needs to be used if the customer's electrical design is not able to meet the specified electrical test requirements using the supplied components and the specified processes. Refer to clause 6(e) in the example of a subcontractor's Conditions of Sale given in Appendix 3B.

When the customer does not define the acceptable thermal and environmental process profiles and their tolerance ranges, the experienced subcontracting firm would state its intentions and send them in writing—if not at time of quote, then at least with the firm's acceptance of the order. Armed with this knowledge, the customer is then equipped to assess the 'fitness for purpose' of the design and normally should expect to be held entirely responsible for it and for the choice of all free-issue materials.

INSURANCE OF FREE-ISSUE ITEMS

Insurance of the customer's property against damage and theft while it is on the subcontractor's premises is a matter that should be agreed on at the contractual stage. If neither the customer's nor the subcontractor's existing policy covers this aspect of their business, the good subcontractor will arrange it, but expect to charge the customer accordingly.

To some extent, the subcontract company's position will be dictated by its order acceptance policy. If a significant proportion of the business is carried out with free-issue items, it would be considered normal practice to have a suitable policy in being as part of the overhead costs.

It is still prudent for the customer to check the terms and the extent of the cover afforded. The maximum sum assured may not always be high enough for a major project.

9.2 The customer sells components to the subcontractor

Some customers, unable to control the supply of free-issue components to their satisfaction, have tried the alternative of selling them to their subcontractors with a margin to cover testing and goods inwards inspection—if applied.

Should the subcontractor be foolish enough to accept this proposition, the great benefits to the customer are:

- Total control over component supply
- An iron fist over the subcontractor's production output
- Complete visibility of the subcontractor's production yield problems and marginal financial gain if they are bad
- The bonus that the carrying of most of the cash flow risk is down to the subcontractor

No more need be said.

10

Checking designs for manufacturability, test and reliability

10.1 Introduction

It is at the design stage that most of the early problems seen on the surface mount production assembly line are born. Few other electronic assembly technologies involve such a large number of variables and are so sensitive to the need to design for manufacture. This fact, taken with the trend towards higher packing density, enhances the importance of getting both the electronic circuit design and the subsequent printed board layout 'right first time'.

The first part of this chapter covers the basic design requirements for surface mount and mixed-technology assemblies and the latter half consists of a checklist which both the quality manager and the purchasing manager may wish to apply before assembly subcontracting occurs.

10.2 Designing for surface mount technology

The skill in designing surface mount assemblies lies in understanding all the variables and in successfully balancing a series of conflicting requirements. The necessity to trade off miniaturization against ease of manufacture is often difficult for design engineers, particularly if they have not received sufficient training and may not have a full appreciation of the production processes.

For success in surface mount technology, all those working in the design team must have an extended knowledge range compared with that needed for conventional through-hole assembly. Indeed, as previously indicated, some managers will not allow any engineer to use a CAD system to layout a surface mount board without first spending at least three months gaining working experience on the assembly shop-floor.

First they must ensure that the assembly is suitable for manufacture by the known means available: it should be easy to inspect, test and rework. Companies that fail to take these criterial into account can waste much money in production trying to find and correct defects.

Long-term reliability should also be considered at an early stage, particularly in professional and military applications. The design thinking must not be restricted to board layout, but should include all the structural and fatigue factors affecting solder joints during board processing, testing, installation and operation.

Also important is an appreciation of the sources, prices and availability of all components specified, as well as the cost structure of the assembly process.

Set against these things, the electrical performance of the circuit has to be preserved, but within practical limits. A brilliant design which cannot be made at the right cost is of little value. Any speed, frequency, power or thermal considerations must be taken into account, and keeping weight to a minimum is becoming increasingly important in many applications.

The primary driving force behind most surface mount designs is high packing density. Designers, sometimes unwisely, may be forced to consider this above all else. Frequently a significant cost reduction is obtainable if the size can be increased by 10 per cent, and when this is put to the project manager, miraculously a little more space is found! This may not always work, but at least it starts the trade-off discussion on the right lines.

The UK Surface Mount Club has a programme in which its executives regularly visit member companies to investigate or audit their progress. The Club finds that almost two-thirds of shop-floor manufacturing problems are design-related, and many organizations lack the necessary close feedback loops to help eliminate them. This criticism has been levelled particularly at design bureaux, but even in vertically integrated companies any lack of early response from the shop-floor usually turns out to be a management problem somewhere along the line.

10.3 Design for manufacture

Each board layout must be undertaken with its assembly process in mind. This can be especially difficult for those who are subcontracting and the advice is to select the assembler and find out his capabilities before starting layout work. In addition to the obvious factors such as board size handling capability, there are many other aspects to consider.

WAVE-SOLDERING

If wave-soldering is to be used, the packing density will be less than that which can be achieved for reflow. Components need more space to prevent shadowing and to maintain high post-soldering yields. They should be oriented in preferred directions in relation to the wave to allow easy wetting of all terminations. The designer must, at this early stage, decide in which direction the board will pass across the wave and make sure that this information reaches the shop-floor.

Some components, such as large PLCCs, are generally not recommended for wave-soldering, although one major European manufacturer of consumer components claims that this is possible. An in-depth knowledge of surface mount component materials, their construction and compatibility with the assembly processes is needed to assure reliability and high yields in production.

Particular care is essential when designing boards to carry large ceramic chip capacitors and high-pincount integrated circuit packages. Depending on the method of construction and the component manufacturer's skills, some of these items are more prone to damage from thermal shock in soldering than others; cf. Chapter 17.

A good grounding in the many 'tricks of the trade' is essential. For example, solder theft pads should be used to prevent shorts on small-outline integrated circuit (SOIC) package leads, and if the adhesive deposition method is not capable of programmable quantity for each location, the design must keep board surface profile variations to a minimum.

REFLOW-SOLDERING

A similar knowledge of the reflow process must be acquired, including an understanding of wetting force and surface tension effects. The only solder available to make the joint is that placed during the screen-printing operation, hence careful design of the screen aperture patterns is necessary. This task should be carried out on the CAD system by the layout engineer rather than be left for the production department to sort out.

Although use of the board outer layer copper pad size is a fair start, at this stage the professional designer will take account of differing chip component termination heights and will provide resist aperture offsets to protect fine track entries. This is done by varying individual aperture sizes and their centre-line positions in relation to the copper areas.

To avoid solder theft, all electrically connected pads and vias and plated-through holes need to be separated by narrow copper track, or at least by strips of integrated resist.

Reflow equipment characteristics such as belt width, transport centring and available temperature profiles are also important in the siting of large components during layout design. Correct formation of solder joints and prevention of damage to components can hinge upon these parameters as well as on process control.

It is important that the pad layout for small components should achieve the four critical balances, i.e.:

1 Balanced thermal mass (copper) at opposite ends of the component, if necessary using short lengths of necked track to achieve thermal isolation
2 Balanced surface tension effects, avoiding unwanted axial and rotational movement
3 Balanced quantity of solder to minimize component movement and tombstoning
4 Balanced copper pad area(s) at opposite ends of chip components. (This is not the same thing as 1 above.)

Generally a higher component packing density can be achieved and assembled more efficiently with reflow technology than with wave-soldering. On the other hand, overall joint defect rates may be slightly worse with reflow owing to the risk of displacement and the absence of mechanical assistance to wetting which is inherent in the solder bath wave motion.

COMPONENT PLACEMENT
When designing the electronic circuit, it is advisable to bear in mind the characteristics of the intended pick-and-place machinery and to try and work within the constraints it imposes. For example, the board should not be designed to carry 60 different component values/types if the maximum number of feeds on the machine is 50. This would require either two passes or having the extra components added manually—both of which are costly.

Making sure that the machine can place as many of the specified component types as possible is another key target. Yet another is to maximize the placement rate by designing the board aspect ratio and circuit step-and-repeat formats correctly. If the machine's optical correction system is needed for fine pitch lead spacings on high-pincount packages, it should be remembered that this may not operate over the entire placement area available and the machine's capability of recognizing the intended fiducial marks should be checked.

VISUAL INSPECTION AND REWORK
Currently the state of the technology is such that many assemblies have to be visually inspected on a 100 per cent basis. Thoughtful design and good spatial perception can ensure that this task can be undertaken efficiently.

Visual access to every solder joint is required, and surface mount components should not be placed underneath or too close to taller through-hole parts. The inspector should have at least a 30° viewing angle from vertical to each joint from at least two sides and preferably three. The layout should be carried out bearing in mind component heights as well as plan areas—unfortunately, most CAD systems do not yet offer help in this essential requirement.

Where practicable, similar component terminations should be grouped together and aligned in one direction to ease the inspector's task. A glance at the relative contributions of inspection time and manufacturing time to product cost will underline the importance of this point (cf. Table 19.2).

In an ideal world rework is unnecessary, but currently few companies can avoid it. To ensure that it is carried out effectively and with minimum loss of solder joint reliability, the layout engineer needs detailed knowledge of the rework tools available on the production line and the techniques to be used for each component type; cf. Fig. 15.1 below.

Inter-component spacings must be sufficient not only to allow access for the tool, but also to ensure that reflow of nearby joints is avoided. Particular attention is paid to wave-soldered components as these need more space around them to enable rotation and shearing of the adhesive before removal.

ELECTRICAL TEST OF POPULATED BOARDS

In-circuit testing should be considered essential for most boards. It is required to check the presence of solder joints as well as to test the function of each component and, above all, to give instant fault location diagnosis.

A test point should be designed in from the start for every node cluster. These points must be sited away from component joints to avoid the problem of test probes temporarily closing up open circuits. They should be distributed evenly across the board to minimize the risk of bow causing harm to solder joints.

To reduce test costs and increase test reliability, double-sided (clamshell) fixturing and fine pitch probing should be avoided whenever possible. One way is to accept a slight increase in board area to enable all test points to be brought on to one side of the board, preferably the low-profile surface. Vias are acceptable as test points if the probe throw can be kept minimal and the fixture bearings are well designed and well made.

RELIABILITY

This subject deserves a book in its own right, but some pointers on a few key issues are given here.

For leadless chip ceramic components mounted directly to organic printed boards, the resulting structural rigidity brings sensitivity to physical distortion of the board. This may arise either through differential thermal expansion of ceramic and the board when temperature variations occur, or from mechanical deformation of the board regardless of temperature.

CHECKING DESIGNS FOR MANUFACTURABILITY 115

Either way, the result can be a fracture in the solder joint or a microcrack within the chip itself.

The same type of failure may follow a single insertion of a bowed board into an incorrectly designed rack mounting or may be due to cyclic stresses applied by attaching the board, for example to a rigid aluminium frame without allowing expansion slots and/or placing spring washers under bolt heads. Further discussion on failure phenomena appears in Chapter 17, together with methods of reducing the risk of their occurrence.

Bow and twist in the printed board can severely affect production yields and reliability; cf. Section 8.2.

Where high packing density brings a greater percentage of copper area on the board, it is advisable to increase its thickness. For example, above Eurocard size, move from 1.6 to 2.4 mm thick material and avoid aspect ratios above 4:1.

On surface mount printed boards, the fine copper tracks needed for high packing densities bring addtional long-term reliability hazards. These come from mechanical, chemical and electrical stresses and are affected by the choice of board manufacturing method, the type of solder resist and the clearance gaps and voltages applied between tracks.

There is no single cure, but again, risks are reduced by awareness of likely stress points and use of intelligent design to minimize them. All tracks less than 0.2 mm wide should be widened before entry to solder pads or protected by resist at the point of entry—especially if dry film is used. All tracks reduced below 0.2 mm width to pass between adjacent solder pads should be widened again immediately afterwards.

Observe conservative gaps between fine tracks where voltages are high or prolonged exposure to damp conditions is likely. On larger multilayer boards, ensure that buried track clearances from adjacent vias and other tracks are never less than 0.3 mm—i.e., do not push the alignment accuracy limits of the average board manufacturing process too far.

10.4 Checking the layout

THE NET LIST

If the CAD system does not have a schematic capture facility or the circuit diagram has not been fully loaded, a copy of each layer of the printed circuit board should be checked against the net list, including all interlayer connection vias and plated-through holes.

BOARD LAYOUT, COMPONENT SELECTION AND EQUIPMENT STRUCTURE

If the CAD system used for printed board layout design has validated inbuilt proximity and orientation rules for components and tracks, a detailed study or formal checking procedure is not usually necessary. However, many systems in use do not include this facility to the required degree. It is therefore essential that a person or a small team outside the design department should check for incorrect or unsuitable layout at the stage before printed board manufacture commences.

There are five basic requirements in these checks.

1 Conformity with layout rules/guidelines

If the company has not developed its own internal guideline documents, procurement of available publications is advised. (Refer to the Bibliography, page 259.)

Because it is impracticable to write down all the factors affecting design quality, and the likelihood is high that even the least stringent rules may need concessions on occasion, an experienced and methodical engineer should be asked to make the conformity checks.

2 Design for manufacture by specified processes and within a cost target

There are no safe short-cuts. Design to suit specific machines and processes is essential, and this is best developed either in-house or in collaboration with an experienced subcontractor who has his own design team and production facilities within the same building.

An important message in this regard is that neither CAD bureaux nor printed board manufacturers may necessarily have sufficiently short feedback loops from an assembler's shop-floor to ensure the best wisdom in all design matters.

The need for close communication between all parties cannot be overemphasized. A heavy cost penalty can be paid for failing to choose the right designer.

Although they are not all strictly quality or reliability matters, some important and obvious checks reflect the foregoing design points:

- Will the board/panel size fit all the available or intended production and test equipment?
- Are all board fiducial marks recognizable as to shape and surface colour by the intended pick-and-place assembly equipment?
- Can all the components be handled by the auto-placement and/or manual placement equipment available? If not, how will they be handled?
- Are all components specified by their manufacturers as suitable for withstanding the intended assembly processes and their respective time–temperature profiles?
- Does the layout contain any problem areas related to the intended method of soldering, e.g. component orientation or shadowing when

CHECKING DESIGNS FOR MANUFACTURABILITY 117

wave-soldering, surface tension, or thermal mass imbalances which may inhibit reflow yields?

3   Mechanical and thermal integrity

For companies operating under IEC(Q) or equivalent national standards, it is the responsibility of quality departments to monitor the performance of the design team to ensure that they are taking all the required early steps to assure product quality and reliability. Important checks for common failings of surface mount circuit designs are:

- Is the board so thin or of such high aspect ratio or with such gross imbalance or copper distribution that significant bow is inevitable? Will this bring associated production yield problems and later reliability risks if it is straightened or repeatedly flexed in operation?
- Bearing in mind the rigidity of some surface mount component attachments to the board, has the assembly structure been designed to survive the mechanical stresses imparted by the intended electrical test fixtures? Alternatively, have the planned test fixtures been designed to avoid stressing the board assembly? Considerable deformation of the board can occur if there is inadequate support during multi-probing.
- Will the structure also survive subsequent assembly of the board into equipment without risk of creating conditions for incipient cracks in solder joints or ceramic chip components?
- If entry and withdrawal from racks with back-wired plug/socket entry is envisaged, will the surface mounted components withstand that?
- Recognizing the differential expansion/contraction occurring between large ceramic leadless components and, for example, FR4 board material, will the structure withstand the cyclic environmental and mechanical stresses experienced by the equipment over its designed life?
- Has the designed life of the assembled board been based on a realistic calendar—including storage, transport, etc., rather than on just operating life?
- Are some of the conventional through-hole components expected to survive any of the surface mount assembly process time–temperature and other combinations? Have these been checked with appropriate suppliers or in their literature?
- Are surface mount connectors used on the printed board? Will their connections to the printed board withstand the likely misuse to which connectors are often subjected? Is there enough space to improve reliability by using through-hole types?

4   Visual inspection and rework aspects

- Has the layout allowed sufficient space around components to ensure

adequate access for intended rework tools to operate without damage to adjacent components?
- Have any small components been placed under large ones? If so, are the large ones socketed to enable easy access to the small ones for visual inspection and rework?
- Are all solder joints visible at a 30° angle to the vertical from at least two out of three directions?

5   Customer/end-user requirements, including product safety

- Check that members of the design department have read and understood the customer/user specifications and have ensured, within the constraints of available knowledge, that the proposed design complies with them and will be reasonably safe when used in the manner and for the purpose for which it has been designed.
- Where appropriate, check that the sales department has made available to the customer suitable warnings and/or instructions for users for products known to require them, e.g. circuits for controlling dangerous processes, life protection or life support systems.

The above product safety-oriented points are considered to be part of the 'duty of care' required in many countries under their consumer protection and health and safety legislation; see Sections 5.7 and 18.5.

11

Specifying the product and defining the basis for quote

11.1 Introduction
This chapter is in the form of a checklist and includes both the obvious and the less obvious items that should be clarified by the purchasing organization—preferably before obtaining quotations—so that like can be compared with like. As far as discussions with potential subcontractors are concerned, inevitably some points will be left until detailed negotiations are in progress, but it is important to consider them well beforehand because one clear measure of a subcontractor's calibre lies in the extent of the questions asked prior to submitting a quotation.

Defining the product with sufficient accuracy is almost half the battle in successful subcontracting. The trouble is that very often it is new product designs that are scheduled for outside manufacture, and if the risks in these are not identified by the customer, the subcontractor can only guess and hope.

11.2 Defining the product
The main points that need to be addressed are well known, but all too often the required degree of professionalism in covering them is absent. Clearly, the total task requires close co-operation between both parties before any contract is signed.

The summary of key headings is as follows:

- Field of application of the end-product
- Function of the product
- Circuit diagram and 'net list'
- Component list
- Printed board specification
- Process method and sequence for which board was designed

- Method of assembling board to equipment
- Operating and storage temperature range of end-product
- Free-issue and subcontractor procurement items
- Electrical test programme
- Electrical test hardware
- Design/type approval testing
- Burn-in and screen testing
- Packing and shipping requirements
- Design tasks and responsibilities, e.g.
    electronic circuit
    thermal
    mechanical
    board layout
    processes and sequence
    testing
- Processes specified by the customer
- Jigs and tools
- Ongoing quality assurance testing
- Responsibility for insuring free-issue goods
- The basis for customer acceptance/rejection at goods inwards
- Delivery quantities and time-scales
- Terms and conditions

Each of these headings is now covered in more detail.

FIELD OF APPLICATION OF THE END-PRODUCT
Stating which of the main market categories is the sales target area is a help to the professional subcontractor who is probably aware of the general needs in each and may provide useful comment. In the context of design and manufacture using surface mount technology and the resulting structural integrity of assemblies, typical categories would be:

- Consumer: 'white', 'brown/black' goods, security, toys, DIY
- Automotive: cold, temperate, hot climates
- Industrial: instrumentation and measurement, control systems
- Medical: life support, other
- Telecommunications: fixed, mobile, hand-held
- Computer/office: fixed, portable, hand-held
- Military: fixed, mobile, hand-held
- Radio communications: fixed, mobile
- Avionics: ground, airborne
- Space: ground, airborne

FUNCTION OF THE PRODUCT

Within the limits of commercial security, it is always helpful in building trust and motivation to state the function of the product. This is no problem when it is a complete item, but even if it is just one board assembly among 20 others in an equipment, its purpose and the overall role of the equipment should be given. The reason is that assembly operators faced with many different boards tend to prefer working with those that they can relate to in terms of their function rather than just knowing the name of the customer.

With this in mind, many subcontractors ask their customers to supply a sample of the end-product so that this can be displayed in a suitable place.

CIRCUIT DIAGRAM AND 'NET LIST'

Nowadays it is considered folly to supply free-issue boards and components without also providing a copy of the electrical circuit diagram and the net list. These documents are a first port of call when things do not go as planned—e.g. wrong components supplied, component polarity reversed, quantities incorrect, board layout incorrect. All components should be identified by a number, and when in-circuit testing is specified, the circuit diagram should show all the test point nodes used.

Aside from enabling the subcontractor to act as a 'long-stop' checker of new designs, this type of information is a prerequisite to meaningful telephone conversations when problems arise.

COMPONENT LIST

Surface mounted component lists should show all chip component termination heights as well as their standard plan sizes (if appropriate), so that print screen apertures, pick-and-place machine pass-through capability and placement programmes can be validated.

In addition to the normal listing of quantities, types and approved suppliers, if the subcontractor is to provide a board layout design, the approximate actual power dissipated in each component should be given so that excessive thermal concentrations can be avoided. In this context, stating the published data on component maximum rated dissipation is useless unless fully utilized.

PRINTED BOARD SPECIFICATION

The level of detail in the board specification needed by the subcontractors will depend on whether they are expected to design and assemble it, procure and assemble it or merely receive and assemble it as a free-issue item. In the latter context, there is also a dependence on who is to carry out the goods inwards check.

To avoid textual repetition, the needs for each respective option are listed in Table 11.1. Most of them are dealt with in greater detail in Chapter 8.

Table 11.1 Printed board procurement statement needs

Detail	Design	Procure	Free Issue
Base material	x	x	x
Board lateral dimensions: existing max/min	x	x	x
preferred max/min	x		
Board thickness: existing max/min	x	x	x
preferred max/min	x		
Number of copper layers		x	x
Copper layer thickness(es) max/min		x	x
Single/double-sided component attachment	x	x	x
Resist type, thickness and positional tolerances		x	x
Solderable surface finish and thickness tolerance		x	x
Solder surface protection material	x	x	x
Preferred/approved suppliers	x	x	x
Circuit operating frequency/switching speed	x	x	x
Bow/twist specification: for auto-assembly	x	x	x
for mounting to equipment	x	x	x
Minimum via/PTH plating thickness	x	x	x
Positional accuracies: of tooling holes	x	x	x
of fiducial marks	x	x	x
Special finishes, gold, nickel, carbon: tolerances of on thickness	x	x	x
tolerances on position	x	x	x
Fine tracks and gaps, minima and tolerances	x	x	x

PROCESS METHOD AND SEQUENCE FOR WHICH THE BOARD WAS DESIGNED

If the board layout has been carried out by the customer or customer's agent, it is vital that the soldering process and assembly sequence for which it was designed should be stated. The necessary information includes the designed orientation of the board on the soldering machine transport mechanism so that surface tension, shadowing and thermal problems which bring extra rework and reduced reliability can be avoided.

METHOD OF ASSEMBLING BOARD TO EQUIPMENT

The customer firm's intended method of assembling the board to its equipment should be defined to the subcontracting firm in case reliability problems are foreseen. These can arise from both mechanical and thermo-mechanical stress sources.

DEFINING THE PRODUCT AND BASIS FOR QUOTE

OPERATING AND STORAGE TEMPERATURE RANGE OF THE END-PRODUCT

These data should be stated so that, if necessary, the subcontractor can adjust the process and limit the rework options. In extreme cases it may be necessary either to change the assembly and solder joint structures or to use alternative board materials.

FREE-ISSUE AND SUBCONTRACTOR PROCUREMENT ITEMS

Where it is intended to split the procurement responsibility between both parties, the respective parts lists should be drawn up. Many customers prefer to purchase items themselves when they possess expensive or the only available electrical test equipment, for example, when there are items such as application-specific integrated circuits (ASICs) or special discrete semiconductor devices to be tested at goods inwards.

ELECTRICAL TEST PROGRAMME

The full electrical test programme and the respective tests to be carried out by the subcontractor and the customer must be defined. The range may include:

- Inter-stage, e.g. after surface mount component assembly and before through-hole components are inserted
- In-circuit component testing after complete assembly
- Functional testing
- Post-burn-in testing
- Testing at temperature and/or voltage extremes
- Testing after mechanical, environmental and/or endurance tests.

Subcontractors will need to know exactly what tests are to be performed after the assembly has left their premises so that they can help to ensure successful results.

ELECTRICAL TEST HARDWARE

Among the decisions to be taken before putting assembly out to subcontractors is the amount of test hardware needed for both bare and populated boards and who is to procure, program, commission and maintain it.

Note that under some national and international factory approval schemes, e.g. IEC-Q and BS 5750, the subcontracting firm is held responsible for ensuring that all the test gear used by it for measurement is calibrated at the required intervals—even if it belongs to the customer.

DESIGN/TYPE APPROVAL TESTING

It is evident that in most instances the reliability of products has been maintained or improved after conversion to surface mount assembly.

However, it is also safe to say that the large number of design and process variables in the technology has severely penalized unwary and untrained practitioners.

Testing to validate the structural integrity of the board assembly and the equipment containing it is normal procedure for new products, but it is especially important with surface mounted assemblies. The range of tests and who is to carry them out should be decided at an early stage so that sufficient extra circuits can be manufactured at the right time.

An additional factor with surface mounting is that the tests should be performed on a product that has been through the exact production process and machinery that will be used in final manufacture. Use of prototypes made by other means will give different results—which may be much better or far worse.

BURN-IN AND SCREEN TESTING

If some form of burn-in or screen testing is required to be performed on a 100 per cent basis, the protocol should be defined and the times and conditions stated.

Often this is just a power-up soak test, either at room temperature or slightly above. Where the product has to survive in hostile environments, then restricted dry heat, damp heat, thermal cycling or mechanical tests are used, but the applied time-scales must represent only a small percentage of their time to destruction under the same conditions, e.g. 2 per cent.

A corollary of the last point is that previous tests to destruction should have been carried out to enable demonstration that the proposed schedule is reasonable.

For burn-in, the amount of shelving, ovens, racks and sockets needed will depend on the output rate and the length of the test in hours or days. These matters must be determined well in advance so that equipment can be specified and ordered.

PACKING AND SHIPPING REQUIREMENTS

Too many companies leave this until the subcontracting firm says it is ready to deliver and then are disappointed when the goods turn up in black anti-static envelopes crammed—or, worse still, rattling around—in a large cardboard box.

A suitable method of transit packing should be decided along with who will be responsible for provisioning it, e.g. any specially shaped mouldings or boxes, warning and other labels. Will the packing be capable of recirculation to and from the customer?

Professional subcontractors usually treat the packing and shipping function as a direct labour area because there can be wide differences in the cost

DEFINING THE PRODUCT AND BASIS FOR QUOTE

of preparing the goods for dispatch. This is done to avoid those customers having easy jobs paying for the difficult and time-consuming ones.

DESIGN TASKS AND RESPONSIBILITIES

Responsibility for the 'design task' and who is the 'design authority' for each of the key subjects should be clearly defined, if not in the 'request for quote', then at least in the contract. An example is given in Table 11.2.

Table 11.2 Design tasks and design authority

	Design task responsibility	Design authority
Electronic circuit design	Customer	Customer
Printed board layout	Subcontractor A*	Customer
Thermal design	Subcontractor A	Customer
Board/component structure	Subcontractor A	Customer
Assembly process sequence	Subcontractor A	Customer
Process parameters	Subcontractor B**	Subcontractor B
Test equipment	Subcontractor C***	Subcontractor C
Protective coating	Customer	Customer

*   For example, a CAD bureau.
**   For example, an assembly only subcontractor.
*** For example, a specialist test equipment design house.

Note that the person doing the board layout is, by definition, also creating the thermal design, selecting the process sequence and the board/component structure, e.g. mixed-technology, double-sided assembly with components on both sides. The consequences of these requirements are discussed in Chapter 12.

PROCESSES SPECIFIED BY THE CUSTOMER

If the customer wishes to specify one or more particular processes, this should be stated in the 'request for quote' together with the precise requirements, for example no wave-soldering of surface mounted semiconductor packages. In preparing its response, the subcontracting firm will need the information to decide whether it is able to fulfil and control the mandatory parameters—and if so, at what cost.

JIGS AND TOOLS

Specialized jigs and tools would be the subject of negotiation only if they are above the normal 'stock in trade' of the subcontractor and are needed to achieve the required price and output rate.

Where a 'part tooling charge' is made, the question of ownership should be agreed as a specific item in the initial contract. This matter is addressed in Chapter 3 and exemplified in the typical quotes given in Appendices 13A and 13B.

ONGOING QUALITY ASSURANCE TESTING

For longer-term contracts, especially those involving products released under military or Capability Approval specification regimes, regular testing of samples from production has to be carried out. These range from non-destructive tests on samples from every batch shipped, e.g. solderability, board flatness, performance at temperature extremes, to quarterly or annual checks such as shock, vibration, acceleration, long-term damp heat, thermal shock and endurance, which are usually destructive and are designed to monitor control over materials and product integrity.

Where such tests are envisaged, if it has not already been established that those on the subcontractor shortlist have the necessary facilities, the matter should be raised in the 'request for quote' so that the subcontractor is aware of the longer-term implications.

RESPONSIBILITY FOR INSURING FREE-ISSUE GOODS

In the case of free issues, their approximate value for insurance purposes should be stated if the subcontracting firm is expected to provide cover while they are on its premises; cf. Chapter 9.

THE BASIS FOR CUSTOMER ACCEPTANCE/REJECTION AT GOODS INWARDS

Both the nature and the extent of the customer's goods inwards and any later inspections of assembled product should be stated. The expected basis for acceptance/rejection at each stage is an important issue because, with mixed responsibilities for various aspects of design and manufacture, the subcontracting firm will not expect to be penalized unfairly for rejection of product when its own contribution is satisfactory.

This can be a very difficult area to manage successfully, especially when the receiving company cannot be expected to make judgements on complex boards until they are integrated with its own equipment and tested over a considerable period.

Also, when a defect is found it can sometimes be very hard to determine the exact liability. Removal of some types of surface mounted components from the board can destroy vital evidence. It may be impossible to say with certainty whether there was a design error, a material defect or a faulty process. Often there will be a mixture of two or three causes.

This subject is dealt with in greater detail in Chapter 17.

DELIVERY QUANTITIES AND TIME-SCALES

The delivery requirements for prototypes, pilot production and full-scale production should be stated, as appropriate. It is helpful to include any pauses forecast for testing and assessment of new designs as this will give the subcontractors confidence that they are associated with a bona fide product.

TERMS AND CONDITIONS

While customers will always rely on their 'Terms and Conditions of Purchase', it is prudent that subcontractors' 'Terms and Conditions of Sale' should be checked for gross incompatibility so that any points to be cleared can be dealt with well before any problems arise. Refer to Chapter 3.

12

Choosing a printed board layout design subcontractor

There are two principal objectives to bear in mind when selecting a printed board design subcontractor. They concern the assessment of the designer himself, and that of the CAD equipment available to him.

Clearly, even the very best CAD system will be wasted if the designer is not competent, though it is possible to obtain a degree of protection in the form of inbuilt software on design rules and other safeguards. In this connection, reference is made to the points contained in the final case history in Chapter 20 concerning some of the risks arising when the design staff are isolated from the realities of the assembly line.

The requirements are best expressed in the form of questions which the customer should raise and the responses that should be looked for.

12.1 The designer

Q1 How many months/years has the designer spent working on a surface mount assembly line?

The answer should be, at a minimum, three months. Any experience shorter than this is unlikely to have provided sufficient breadth of training. Certainly if the answer is 'None', then the best advice is to go elsewhere.

Q2 How many surface mount and mixed-technology board layouts has the designer made?

Q3 Of what order of complexity were they? Can they be viewed?

The replies to questions 2 and 3 should provide assurance that the designer has worked on surface mount circuits containing component counts and pincounts comparable to those under consideration.

Q4 How many of the surface mount and mixed-technology boards have been digital and how many analogue?

The layout background needed in these two circuit categories is very different. A high-frequency analogue circuit should not be subcontracted to a designer experienced solely in simple digital circuitry—the knowledge gap will be too great.

Q5 How many of the surface mount or mixed-technology boards have been multilayer? (meaning four layers or more)

Experience shows that the opportunity for errors of judgement in placement of components and tracks increases rapidly with the number of layers. The need for higher intellectual capacity also increases rapidly with each extra layer involved.

The same type of questioning is appropriate for the special knowledge requirements of power-dissipating and very high-frequency circuits.

An important point being made above is that it is insufficient to go to a design house and talk only to the management or the chief designer. The assessment must be based on an interrogation of the person who will actually perform the layout task. This is because, once a complex layout has been made, it is extremely difficult for a more senior person who has not been involved in the vast number of individual decisions to check them all and validate the overall design.

Further means of assessment of the individual designer's calibre is given later in this chapter, i.e. by the level of questions asked of the customer.

12.2 The CAD system

The evaluation of a CAD system and its software is outside the scope of this book. Most design subcontractors purchase their systems with the limited objective of carrying out the layout tasks. Some do offer the ability to validate electronic circuit designs by worst-case or sensitivity analysis and schematic capture, but these are few and far between and remain mainly the prerogative of large OEMs.

Some of the general points made here concern the subcontract houses' capability to assist the customer in ways that are outside the narrow design sphere. A vital first question on the software side is

'What layout design rules are to be used?'

If the response is 'Ours', meaning those of the subcontractor, then there are three supplementaries:

- Where did they originate?
- How have they been validated in production?
- Are there several other customers that can be approached for an opinion on this point?

In following up the last question, it is important to talk to the production management in the selected customer-firms rather than to the design people.

Moving to the main technical requirements from the CAD software, we ask:

Does the system output capability cover the following:

- Net list
- Parts list
- Sufficient active layers to cover the multilayer and other needs
- Stage-by-stage assembly drawings if required
- Compatible drilling and routing control tapes/floppy disks for the pre-selected printed board manufacturer
- Excess thermal density warning system
- Copper balance indication
- Ohmic resistance between any two points—important for large boards with fine tracks
- Impedance of designated tracks to ground or power planes
- Mutual inductance/capacitance between any selected adjacent tracks
- Off-line pick-and-place and/or glue-spotting machine programme capability. Probably this means using one of the standard interfaces available.
- Metric-to-inch conversion to the required number of decimal places. This is vital if the CAD system is based on inch measurements (e.g. has been made in the USA) and the circuit contains fine pitch integrated circuits made in Europe or the Far East. Refer to the postscript at the end of Chapter 20.

Subcontractors' outputs which can materially assist the customer's production planning, and which would be expected from any professional OEM's in-house CAD system, are:

- Pincounts for each board surface. This data is needed to calculate standard rework and visual inspection times and costs based on forecast p.p.m. defect rates and work-study synthetic standard times.
- Number of different component feeds required for both pick-and-place and auto-insertion machines. This data is needed for selecting the assembly subcontractor and estimating the machine occupancy and loading/unloading times for machine set-up. Different machines use up different numbers of 8 mm feeder slots for the same component types.

12.3 The job to be done

Questions that a competent layout designer who is used to the rigours of design using surface mount technology should ask of the customer are given

CHOOSING A LAYOUT DESIGN SUBCONTRACTOR   131

below. All affect the layout and should therefore be asked and answered before the design starts. They are not put in any order of priority, but, because they are structured to protect the interests of the customer by carrying out the 'duty of care', failure to ask each one must be considered a point against the designer candidate:

- What soldering process types are available at the in-house or subcontractor facility and can be used in assembly, e.g. wave, infra-red reflow, vapour phase reflow, thermode (hot bar) reflow?

This information controls panel/board shapes and sizes as well as component footprint pad sizes, orientations and proximities.

- If infra-red or vapour phase reflow-soldering is intended, is the layout of the solder paste printing screen or stencil to be included in the design task? (It should be!) If so, is the printing machine to be used capable of so-called "on-contact" as well as "off-contact" mode, and can screen thickness and intended solder paste metal content percentage be specified?

'On-contact' machines are essential for the fine line stencil printing required for fine pitch and TAB device packages. Screen thickness and paste metal content data are needed to decide the printing screen aperture sizes and hence to control the paste quantity delivered at each footprint pad. Printing paste material may not be a suitable method of applying the solder for lead pitches below 0.4 mm.

- What soldering processes can be used with the specified components?
- What is the required board thickness?

This can affect board flatness and multilayering techniques applied.

- What solder resist material is available/preferred from the intended board supplier(s), and what thickness is prescribed—if any?

The choice of resist type controls the designed clearances around solder pads and the proximities achieved for components, test pads and via holes. Its thickness affects the ability to use wave-soldering, to tent via holes and to perform successful rework operations on integrated circuit packages with fine pitch lead spacings.

- What assembly structural options are suitable for the equipment, e.g. single or double-sided, mixed-technology, preferred sides for surface mount and through-hole components?

In some cases there is no choice, and the board layout is limited to the required type.

- How is the board to be mounted into the equipment?

This question shows that the designer understands the reliability problems that can arise from mechanical and thermo-mechanical stresses applied to the board and to surface mounted components. They can be harmful while it is being mounted into the equipment and during its subsequent lifetime operation.

- Is there a preferred board panel size for the chosen production line?

Note that this point can be properly answered only when the maximum panel size capability of the selected assembly line has been established. It requires checking in respect of screen printing, adhesive placement, component placement, adhesive curing, component insertion, soldering, cleaning and electrical test equipment.

- What screen printing equipment is available?
- What board location method/dowel hole method is required?
- If optical correction is necessary, e.g. for fine pitch work, what fiducial marks' configurations are recognized by the printing machine, what surface finish for the marks is needed, and where should they be located on the panel?
- Are manufacturer's drawings available for all surface mounted components?

For example, it is not sufficient merely to state standard chip plan sizes (0805, 1206, etc.) when reflow-soldering is intended because the design of screen-printing apertures is dependent on the thickness of capacitor chips as well as their plan area. Also, the variation in lead dimensions for alleged 'standard' Japanese Quadpacks is far too wide for high yield on soldering, and the layout must be tailored to the particular manufacturer's drawings.

- What are the characteristics of the selected pick-and-place machine?

Things the designer must know before starting work include:

— Does the machine have pass-through capability and, if so, what are the methods of board location that must be specified as part of the layout drawing?
— If the machine has a simple platen, what are the alignment dowel pin sizes and positions?
— What are the machine Go/No-go arears on the board in the case of both tracks and components?
— What are maximum above- and below-board component heights that the machine can handle?

- If optical correction is necessary, e.g. for fine pitch work,
  — What fiducial marks' configurations are recognized by the pick-and-place machine?

CHOOSING A LAYOUT DESIGN SUBCONTRACTOR   133

— What surface finish for the marks is needed (e.g. bare copper, solder-coated copper)?
— Where should they be located on the panel?
— Can the same marks be used for the printer and the pick-and-place machine?
— Are there any restrictions on the location of large, fine lead pitch or TAB components which require individual local fiducial marks?

- If auto-insertion machines are to be used, what are the clearances that must be allowed for insertion jaws (above the board) and clinching tools (below the board) to ensure that surface mount components already assembled are not damaged?
- What range of surface mount rework tools is available on the assembly line, and what clearances are needed between components to ensure their correct application?

This requirement is forgotten time and time again by designers, and circuit reliability is reduced through production staff having to use the wrong rework tools. The data given in Table 15.1 can be used to check that the right equipment is available to suit all the components on the board.

- What types of visual inspection are to be used and, for example, what throat depth is available on projectors, microscopes and other equipment, including automatic inspection machines?

These factors can affect component proximity and board size.

- What electrical in-circuit and functional test equipment will be available, and what method of holding probe sets to the board is to be used, e.g. for single-sided or clamshell fixtures? Should space on the board be allowed for support pillar locations to prevent the board bowing during test?

12.4 Conclusion

From study of the questions asked in this chapter, readers will have little difficulty in realizing that, to make a reliable and cost-effective design, the board layout engineer must have available a large amount of information on the machinery, processes and components that are to be used in assembly.

This fact provides the educated customer with a unique opportunity to assess the subcontract designer's experience and calibre and it is unwise not to take advantage of it. The best subcontractors will welcome the approach.

13

Choosing an assembly subcontractor

13.1 Financial aspects

Investigating the financial status of a subcontractor should precede all other evaluations. Apart from published accounts and material supplied by organizations such as Dunn & Bradstreet, most subcontractors are willing to provide information that can be used to assess integrity, e.g. creditworthiness, names of backers, bank references, etc. These matters are much easier and less costly if both sides are willing to co-operate by exchanging such data (cf. Chapter 14).

It may be harder to be certain that cash flow requirements can be met should things not go exactly to plan—especially if the company has been expanding rapidly. This is the most common problem facing smaller subcontractors.

Where a subcontractor is a division of a major group, its individual profit and loss data are unlikely to be published separately and may not be forthcoming. Balancing this is the nominal assurance that comes from doing business with a larger organization. The word 'nominal' is used because, today, large companies and their divisions appear to be no more secure than small ones.

The size of a subcontractor's order book is not a useful yardstick unless there is precise information on the time-scale over which deliveries are scheduled. The definition of what constitutes a firm order can often be the subject of optimism; for example, a large annual order placed in January, with monthly call-off arrangements and cancellation without penalty at three months' notice, is not a firm commitment beyond, say, May, but it may be quoted in the books as though it were valid for the year.

An important and sometimes 'tell-tale' point is whether the subcontractor enters all overmakes into his finished goods stores at zero value. This is generally regarded as sound practice, but if it is found not to be so, it could represent a sign of previous or present financial problems. The key issue is whether the entry at sales value is automatic or discretionary and, if the latter, at whose discretion? One might have more confidence if it is the MD's decision rather than that of the sales manager.

A demand for a proportion of the prototype price to be paid with order placement, or for all the prototype materials to be covered by an 'up-front' payment, is more likely to be a sign of a well-run company than an indicator of financial difficulty.

Reluctance by a subcontractor to give the names of large companies which could give references should not necessarily be taken as a negative sign. Many such companies prefer to restrict public knowledge that they are using subcontract assembly houses and the professional company will seek to honour this concept to the letter. If it does not maintain commercial security in all matters, an assembly subcontracting firm will soon be out of business.

13.2 The subcontractor's team

Apart from technical points discussed below, the customer visiting a subcontractor for the first time will want to review the calibre of the team that may be doing the work. Upon leaving, the customer will ask, 'Whom did I meet and what did I think of them?'

The good subcontracting firm will have gone out of its way to call in and introduce all the principal members of its team, including those heading the purchasing, technical, quality, production/materials control, shop-floor supervision and sales functions.

13.3 Defining the capability and capacity requirements

Many large OEMs have a policy on subcontracting that work for them in any one company should not exceed 20 per cent of that company's total capacity. Clearly, an ability to estimate the true capacity on an independent basis is essential.

This section is intended to provide guidelines for doing key parts of the job; but before attempting to assess a subcontractor it is necessary to have a clear idea of the magnitude of the work to be subcontracted.

There are several yardsticks that can be applied and usually more than one is needed. The first is component count. What are the forecast respective rates of usage of surface mount and through-hole components? These figures must then be related to the auto-placement and auto-insertion capacity of the subcontractor's machines.

The effective capacity of surface mount placement machines is almost invariably overstated. This is because of the industry's standard practice in brochures of assuming very short placement traversing distances and/or the use of maximum operating speed figures which may not be practicable for medium and large sizes of component.

If the subcontracting firm is into regular high-volume production with comparatively few different circuit types, it will not have to spend too much time programming and setting up the machines and a figure of 50–60 per cent of the machine manufacturer's claims should be assumed. This allows for programming and set-up, loading/unloading and other downtime factors as well as the reduced-speed requirement. Typically, a pick-and-place machine rated at 14 000 placements per hour would be assumed for costing purposes by the subcontractor, to operate at around 7500.

It must be said that one cannot directly compare these figures with what can be achieved by an OEM with a single product line. In such cases the utilization figure could be in the 75–80 per cent region. If programming and resetting of the machines occurs on a daily basis, then the subcontractor's figure should be assumed to drop to around 40–45 per cent.

There are two further factors related to components. One is the number of feeds on the pick-and-place machine. If there are 80 different surface mounted components in the customer's circuit and the subcontractor's machine has a capacity limitation of 60 feeds, then either there will have to be a second pass through the same or another machine—with consequent resetting and paste deterioration problems—or the extra 20 components will need to be handled manually. In either case there will be a yield/cost penalty and a risk of reduced reliability.

Here again some manufacturer's brochures are misleading, in that they quote the capacity in terms of the number of 8 mm tape feeders. The reality will depend on the number and variety of feeders, and these will vary in their proportion of machine capacity taken up according to tape and stick widths. A 56 mm wide tape carrying Quadpack integrated circuits will probably occupy four to six equivalent 8 mm tape positions, and a typical circuit would reduce the actual feeder capacity to 50–60 per cent of the brochure maximum claim—sometimes even less.

The second relates to the ability of the machine to handle all the component types that are to appear on the printed board. Some machines are not suitable for placing very small and very large types; others can handle only small components but do so at very high speed. Some larger subcontractors overcome this problem by having several types of machine in line so that the board makes only one pass but does so through two linked machines via a mechanized transfer system.

If the board contains components needing optical correction to obtain the necessary placement accuracy, it should be checked whether the total placement area on the machine platen is available for such components. Usually there is a restriction which must be taken into account during the board layout stage.

Other aspects of placement machines needing consideration include the maximum printed board size and the maximum component heights above and below the board that can be handled, especially on pass-through machines.

Undue attention is often focused on placement capacity when in fact there are other resources that are also limiting factors. The size of the materials store can be a serious restriction for a small subcontractor seeking rapid growth. Moving down the line, the capacity for depositing solder paste for reflow methods and adhesives for wave-soldering should be compared with the customer's needs. In both cases, a syringe type dispensing system is more likely to be a bottleneck than a screen-printing approach.

Soldering capacity is usually found to be much greater than can be handled by the above-mentioned equipment, but it must be checked that the process type and the direction of entry into the machine for which the board has been designed are within the subcontractor's capability, for example on belt width for an infra-red reflow unit. Subsequent cleaning equipment may be a limitation, not on its capacity to receive the board size, but in its ability to handle the throughput level without loss of cleaning quality.

Other matters needing investigation are on the electrical test gear side (see Chapter 15) and, last but not least, the size of the trained workforce. While it may be claimed that the labour market in the area allows rapid recruitment, training in surface mount assembly operations is known to require much longer periods than for conventional assembly; a relative time-scale factor of three or four is appropriate.

13.4 Questions asked by professional assembly subcontractors

The questions that subcontractors ask of their potential customers give a strong indication of their degree of professionalism. Generally speaking, the fewer the questions asked before quoting, the more worried the customer should be. But this is an oversimplification, because the quality of the questions is as important as the quantity.

To give some indication of the scale of thinking to be expected from a first class subcontractor, sample questions in various subject categories are given in Chapter 14. For every question that is not asked (unsolicited) by the assembly subcontracting firm, a point should be deducted from its scoresheet!

13.5 Reviewing quotations

Apart from the obvious matters concerning price and delivery, key questions to ask when reviewing quotations for surface mount assemblies include:

- Have the deliverables for each stage been properly defined?

- Has the subcontractor entered reasonable technical provisos?
- Are the prices for each stage defined?
- Has the quality assessment side been addressed in a professional manner?
- Have any qualifications been entered on reliability issues?
- Has provision been made for testing the assembly structure?
- Does the electrical test programme meet the customer's stated needs?
- Are the respective design and processing responsibilities defined?

13.6 Sample quotations

The sample quotations that follow in Appendices 13A and 13B are to and from fictitious companies for fictitious products. They are not exhaustive and do not cover all situations, but their purpose is to form a checklist of possible items to consider when preparing or comparing quotations.

While the customer should draw confidence from the depth of detail covered in a subcontractor's response, it may well be that time-scales permit only the briefest of facsimiles giving prices and deliveries. From the subcontractor's viewpoint, customers that base their decision solely on this type of scant information are probably not worth having.

Professional subcontractors will always follow up quickly with a full quotation.

Appendix 13A Example Quotation for Prototype Assembly

To: Psitronics Ltd

Ref: Your Request for Quote No. ____, dated ____
      Circuit Type No. ____

We are pleased to offer our quotation for the assembly and inspection of prototypes of the above Circuit Type No. as follows:

1 Prices

Prototypes in accordance with your layout drawing No. ____
and specification No. ____ :

50 at £195.50 each	Total £ 9 775.00
or 100 at £165.80 each	or £16 580.00

Part tooling and part design charge items:

Conversion of Psitronics board layout to step and repeat pattern	£ 250.00
Photomasters for manufacture of boards and solder print screen	£ 150.00
Printed board drill and rout disc	£ 35.00
Bareboard test jig	£1 500.00
Bareboard test software program	£1 750.00
Purchase of print screen	£ 65.00
Printer location jig	£ 250.00
Pick and place program	£ 400.00
	£4 400.00

2 Basis for quote

2.1 Free-issue items: All semiconductor devices to be provided free issue by Psitronics

The above prices are based on the following assumptions concerning free issues:

(a) That all the surface mounted components supplied are suitable for

infra-red reflow soldering. Data on the intended time–temperature profile for this operation can be made available for your suppliers if necessary.
(b) That the component bodies are suitable for auto-placement using our Dynapert MPS 318 pick-and-place machine and cleaning after soldering by immersion in Dupont 'Axarel' using ultrasonics.
(c) That the solderability of the component terminations (and the printed circuit boards) supplied are 'fit for purpose' when using reflow soldering.
(d) That sufficient quantities of all the parts needed to manufacture the 50 prototypes reach us 8 weeks prior to the agreed delivery date. In this case we suggest a yield allowance of 5% extra parts is appropriate—this figure will reduce after the prototype phase.
(e) It is preferred for all free issue components that the date code of manufacture of the product is available either marked on the component body or on the primary packaging.

2.2 Procurement of non-free-issue items
(a) All remaining components and the printed board to be purchased by Subcon using Psitronics photomasters, parts lists and, where specified by Psitronics, named sources. In the latter case, we assume that Psitronics will check that the components specified are suitable for the processes as stated in 2.1(a), (b) and (c) above.
(b) The solder used will be 60% tin, 38% lead, 2% silver.
The flux type will be rosin, mildly active (RMA).
The cleaning fluid used will be semi-aqueous, Dupont 'Axarel'.
(c) At the end of this contract, all items for which payment has been received will be returned to Psitronics.

2.3 Board layout
The board will be of FR4 material and purchased to our standard printed board specification (copy attached). The solder resist will be of a photo-imageable wet film type, 30–40 microns thick, bonded direct on oxidized copper. The solder coating will be hot-air-levelled tin–lead of thickness 15–50 microns. This will give a useful reflow solderability life after manufacture of 3 months under correct storage conditions.

Our price assumes that the suggestions to improve manufacturability made by our engineers concerning the board layout will be actioned; in particular:

CHOOSING AN ASSEMBLY SUBCONTRACTOR

- R42 to be moved away from IC4 to give a minimum inter-pad clearance of 2.0 mm so that the correct rework tool can be applied to IC4 if necessary.
- C2 is an X7R ceramic chip capacitor of size 2220. On reliability grounds we strongly recommend that it be replaced by a conventional axial wire lead version. The 2220 size is unlikely to survive the recommended type approval cycling tests on FR4 board material.
- C22 to be removed from beneath IC12 and located elsewhere to allow visual inspection of C22 and avoid having to remove IC12 if C22 is faulty after soldering. If this is not possible, IC12 should be socketed.
- A neck of copper track 0.3 mm wide and 1.0 mm long should separate the footprint pad for C11 and the via which is at present integral with the pad. This is needed to prevent solder theft regularly causing a dry joint at one end of C11.
- C43 (100 nF) ceramic chip capacitor in 1206 size should be replaced by the same value in 1210 size and the printed board footprint be altered to suit. As discussed with your engineers, this action is recommended on reliability grounds.
- Your photomaster for the resist has been dimensioned as if for use with conventional screen printed wet film resist. This is not accurate enough for the closely spaced layout you require. The apertures should be reduced in size to give a 0.13 mm clearance all round the edges of each copper footprint pad. This will align with your stated requirements and our recommendations on printed board procurement using photo-imageable wet film resist and is essential to the achievement of our budgeted yield and price.

3 Inspection and test

In accordance with your instructions, all prototypes will be shipped without any electrical test but with visual inspection of all solder joints.

We will use our best endeavours to ensure that all solder joints are satisfactory, but without the addition of electrical testing, this cannot be guaranteed to a high confidence level.

We note that some of the passive chip components have not been provided with electrical test access nodes and cannot therefore be probe-tested individually. We recommend that suitably located probe test pads are incorporated in the design before the production phase starts. These should be 1.0 mm diameter where possible, but can be reduced to 0.6 mm if necessary, and by screen printing enough solder paste on them to ensure that the reflowed mound is above the resist level, a 'crown' probe can be used.

Revised solder paste print screens will then be required in addition to the modified printed board layout.

4 Quality and reliability

Our visual inspection standards are based on IEC Specification No.⎯⎯⎯.

You have advised that the circuit is to be used in street traffic light control systems in central Europe and that they will be housed in unheated metal boxes at pavement level and not operating at night between the hours of 1.0 a.m. and 5.0 a.m.

If the intended product life is 5 years (1827 days), to check the fitness for purpose of the structural design we recommend that at least ten of the prototype assemblies should be submitted to type approval testing. This could consist of temperature cycling tests on assembled boards taking them between $-30\,°C$ and $+20\,°C$ for 500 cycles (winter night simulation) followed by testing for a further 500 cycles between $+20\,°C$ and $+70\,°C$ (summer day simulation). We also recommend that Psitronics seeks additional expert advice on this issue. We would be pleased to quote for carrying out these tests.

5 Psitronics goods inwards acceptance

This quotation is made on the understanding that Psitronics will accept all functionally good circuits and return all defective ones together with a precise diagnosis of which particular components on each circuit are not satisfactory. We would welcome brief written notes with each reject board and, wherever possible, that the defects are marked with adhesive arrows.

6 Delivery

Subject to agreement on items in paragraphs 2 and 3 above, we can deliver the first 10 within 6 weeks of receipt of all components and the remainder within the subsequent 3 weeks.

However, as the circuit has not previously been made using surface mount technology, we recommend a pause for evaluation of the first 10 before proceding with the remainder.

7 Design authority

Our quotation assumes:

(a) That Psitronics are the design authority for all electrical and mechanical aspects of the product.
(b) That Subcon are responsible for controlling the process sequence—as defined by your layout design requirement—and for controlling the quality of solder joints made thereby and the cleanliness of the assembly.

8 Control of change

We assume that Psitronics will notify all required changes in writing and that no action will be taken by Subcon to implement them until such notification has been received from an authorized person nominated by Psitronics.

If required, Subcon are prepared to notify Psitronics of all major changes in processes and materials that are within the agreed responsibility of Subcon and applicable to the product, prior to implementation. A list of notifiable changes and a price for this extra service are available on request.

9 Production quantities

Please note, this paragraph is not a firm quotation, but an estimate.

Subject to the implementation of the above recommended design changes and other manufacturability improvements which may emerge from prototype manufacture and trials, the same procurement arrangements as for prototypes, the provision of suitable production tooling and in-circuit test rig and software, our budgetary estimates for quantity prices are as follows:

For 10 000 off delivered at a rate rising to 500/week	£31.45 each
For 50 000 off delivered at a rate rising to 1000/week	£28.35 each

The above price estimates do not include electrical test of the assembled circuit as it is not possible at this stage to define the length of time required for it.

Assuming only minor layout changes, the additional part-tooling charges payable to achieve the above prices would be approximately:

Revised step and repeat layout	£ 250.00
Revised photomasters for boards and solder print screen	£ 150.00
Revised printed board drill and rout disc	£ 35.00
Revisions to bareboard test jig	£ 350.00
Revisions to bareboard test software program	£ 400.00
Purchase of revised solder paste print screen	£ 65.00
Revised pick and place program	£ 400.00
Handling jigs	£ 850.00
	£2 500.00

We believe that our Gen-Rad ATE has sufficient nodes to provide satisfactory in-circuit testing of components on a 100% basis provided the nodes recommended in 3 above are incorporated on the printed board. If you require us to perform this test, the additional tooling charges are estimated as follows:

ATE test rig £2 600.00
ATE software program £3 250.00

£5 850.00

10 Validity of quotation
This quotation is valid for 30 days from the date given above and subject to Subcon Ltd's Terms and Conditions of Sale.

11 Value added tax
The prices given in this quotation do not include value added tax which may be payable in addition, depending on the services given and the location of the customer's country.

12 Payment
12.1 Payment terms for product shipped to customers: 30 days net from Date of Invoice.

12.2 Payment for part design, part tooling (including test hardware and software) or other tooling charges: 50% with Order, the remainder within 30 days of dispatch of first prototype or the first production batch, as applicable.

Appendix 13B Example quotation for production quantities

To: Psitronics Ltd

Ref: Your Request for Quote No. ____, dated ____
      Circuit Type No. ____

We are pleased to offer our quotation for the assembly and inspection of production quantities of the above Circuit Type No. as follows:

1 Prices

1.1 Product
Circuit assemblies in accordance with your layout drawing No. ____ and specification No. ____

10 000 off level

First 500	each £39.72
Next 1000	each £37.20
Remainder	each £36.33

[*Note that by quoting in three stages, the subcontractor has given himself a small extra degree of flexibility for further negotiation without lowering the remainder price!*]

25 000 off level	each £35.00
50 000 off level	each £34.00

1.2 Tooling charges
(a) Part tooling and part design charge items

Revised step and repeat layout design	£ 275.00
Revised photomasters for boards and solder print screen	£ 150.00
Revised printed board drill and rout disc	£ 45.00
Revisions to bareboard test jig	£ 300.00
Revisions to bareboard test software program	£ 400.00
Purchase of revised solder paste print screen	£ 65.00

Revised pick and place program £ 400.00
ATE test hardware (1454 node probe test rig) £2 950.00
ATE software program £3 350.00

£7 950.00

(b) Additional part tooling charges

For 2000/week output rate: handling jigs £ 950.00
For 2500/week output rate: handling jigs £1 350.00
For 3000/week output rate: handling jigs £1 700.00

Other jigs and tools needed for production will be funded by Subcon.

1.3 Retrospective price adjustments

In the event that the quantities ordered and shipped from continuous production within any six-month period exceed or fall below the respective 10 000 off, 25 000 off and 50 000 off levels, the relevant price levels will apply retrospectively, i.e.

10 000 to 24 999 each £36.33
25 000 to 49 999 each £35.00
50 000 and above each £34.00

2 Basis for quote

2.1 Free-issue items

ICs 11 and 14 (ASICs) to be tested electrically and provided free issue by Psitronics.

The above prices are based on the following assumptions in respect of free issue items:

(a) That all the surface mounted components supplied are suitable for infra-red reflow soldering. Data on the intended time–temperature profile for this operation can be made available for your suppliers if necessary.

In the case of the ASICs, you have decided to do the 100% electrical testing yourselves before sending them to us and will therefore have removed them from their sealed dry packages. Unless they are resealed by you to the manufacturer's dry gas standard or put in sealed bags with fresh dry desiccant, we will not be able to guarantee their freedom from the 'popcorn effect' after mounting without pre-baking them. Apart from the extra cost we would have to pass on to you, this latter pro-

CHOOSING AN ASSEMBLY SUBCONTRACTOR   147

cess reduces lead solderability. Our price assumes that you will reseal them and that we will not have to do the baking.

(b) That the component bodies are suitable for auto-placement using our Dynapert MPS 318 pick and place machine and cleaning after soldering by immersion in Dupont 'Axarel' using ultrasonics.

(c) That the solderability of the component terminations (and the printed circuit boards) supplied are 'fit for purpose' when using reflow soldering. Our price includes the cost of solderability testing at our goods inwards when required.

Note: A price reduction of 0.1 pence per component is offered for each device whose date code of manufacture is less than 3 months prior to our receiving it. This offer is contingent on Psitronics identifying such components in advance of shipment.

It is recommended, for all free-issue components, that Psitronics ask their suppliers to ensure that the date code of manufacture of the product is available marked either on the component body or on the primary packaging. In high-volume production, the date of reeling (if different from the date of component manufacture) should also be marked on each reel.

(d) That sufficient quantities of all the ASICs needed to assemble the first 1500 circuits reach us 6 weeks prior to the agreed initial delivery date and thereafter 4 weeks prior to the subsequent delivery dates. In this case we require a yield allowance of 1.5% extra parts on the first 1500 reducting to 1% for the remaining part of the order.

(e) Our price assumes that Psitronics will insure all free-issue items for loss or damage while they are in transit to or from or at our premises. We are prepared to quote for such cover if it is not inherent in Psitronics' existing insurance arrangements.

(f) Normally we delay commencment of a production batch run until all components requiring auto-placement or auto-insertion are available. In the event of a small shortage, we would limit the batch size to suit the available number of complete kits of parts. If the shortage arises from non-delivery of free-issue parts, we reserve the right to adjust the price of the relevant batch or effect a delayed delivery, depending on Psitronics' priorities and the Subcon forward output schedule.

(g) At the end of this contract, all surplus free-issue materials supplied by Psitronics will be returned.

2.2   *Procurement of non-free issue items*

(a) All remaining components and the printed boards to be purchased by Subcon using Psitronics photomasters, parts lists and, where specified by Psitronics, named sources. In the latter case, we assume that

Psitronics will check that the components specified are suitable for the processes as stated in 2.1(a), (b) and (c) above.
(b) The solder used will be 60% tin, 38% lead, 2% silver.
The flux type will be rosin, mildly active (RMA).
The cleaning fluid used will be semi-aqueous, Dupont 'Axarel'.
(c) At the end of this contract, all surplus components and boards for which payment has been received will be returned to Psitronics.

2.3 Board layout

The board will be of FR4 material and purchased to our standard printed board specification (copy attached). The solder resist will be of a photo-imageable wet film type, 30–40 microns thick, bonded direct on oxidized copper. The solder coating will be hot-air-levelled tin–lead of thickness 15–50 microns. This will give a useful reflow solderability life after manufacture of 2 months under correct storage conditions.

Our price assumes that the suggestions to improve manufacturability made by our engineers concerning the board layout have been actioned; in particular:

— The solder resist apertures for the ASIC leads in their revised packages are still designed around each individual pad. At the new 0.635 mm lead spacing, the strips of resist between pads are very slim, and if rework is necessary there is a strong likelihood that they will lift away from the board and curl over the pads. This causes prolonged delicate and highly skilled work with a scalpel, with consequent risk of further damage to tracks.

   We recommend that the single slot method is used, with extended pads and resist overlapping the ends of the pads.

— The solder paste printing method used in production will require a stencil instead of a mesh screen. We note that the aperture sizes in the photomaster appear to have remained designed for the mesh screen which is 0.4 mm thick. The recommended stencil thickness is 0.2 mm, and although the aperture will contain no wire mesh, there will still be insufficient solder paste deposited. The lengths of the resist apertures should be increased by 20%. This could mean that the copper pads may also require a small extension.

— We note that our recommendation to increase the type size of C43 from 1206 to 1210 has not been accepted. The greater number of metallized layers needed to obtain the 100 nF value in the smaller size increases proneness to thermal shock during soldering. We always endeavour to minimize this, but ask designers to avoid unnecessary risks.

3. Inspection and test

3.1 Incoming components and printed boards will receive goods inwards inspection on a batch sampling basis in accordance with the requirements of Subcon Ltd's Quality Manual.

3.2 Assembled boards will receive visual inspection in accordance with the Subcon Ltd Quality Manual. These range from 100% of all joints to appropriate sampling levels, depending on printed boards, components and processes.

3.3 All circuits will receive an in-circuit electrical test.

4. Quality and reliability

Our visual inspection standards are based on IEC Specification No._____.

As we have had prior experience of your circuit and are carrying out an in-circuit test on a 100% basis, an initial target for post-final inspection solder joint defect rate maximum of 10 p.p.m. can be set. To achieve better than this it will be necessary to devise additional test methods, e.g. burn-in or soak tests, vibration, laser/infra-red inspection, X-ray.

We would be pleased to quote for providing an estimate of 'MTBF' for the assembled board based on the appropriate MIL STD.

5 Psitronics goods inwards acceptance

This quotation is made on the understanding that Psitronics will accept all functionally good circuits and, at their discretion, return all defective ones together with a precise diagnosis of which particular component on each circuit is not satisfactory.

6 Delivery

Delivery ex-works commencing in Week 1 as follows:

6.1 10 000 off

Weeks	1	2	3	4	5	6	7	8	9
Shipments	500	750	1000	1250	1500	2000	2000	1000	
Cumulative		1250	2250	3500	5000	7000	9000	10000	

6.2 25 000 off

to Week 14

| Shipments | 500 | 750 | 1000 | 1250 | 1500 | 1750 | 2000 | 2250 | 2500 | ⟶ |
| Cumulative | | 1250 | 2250 | 3500 | 5000 | 6750 | 8750 | 10000 | 12500 | |

6.3 50 000 off

Weeks	1	2	3	4	5	6	7	8	9	10	11
											to Week 23
Shipments	500	750	1000	1250	1500	1750	2000	2250	2500	2750	3000 →
Cumulative		1250	2250	3500	5000	6750	8750	11000	13500	16250	19250

Note: Week nos. are from date of first shipment.

To meet the above schedule based on your stated requirements, we will need your order 14 weeks prior to first shipment.

7 Design authority
Our quotation assumes:

7.1 That Psitronics are the design authority for all electrical and mechanical aspects of the product.
7.2 That Subcon are responsible for controlling the process sequence—as definded by our layout design requirement—and for controlling the quality of solder joints made thereby and the board cleanliness.

8 Control of change
The above delivery schedule assumes no changes in design after the placement of the Order.

We assume that Psitronics will notify all required changes in writing and that no action will be taken by Subcon to implement them until such notification has been received signed by an authorized person nominated by Psitronics.

If required, Subcon are prepared to notify Psitronics of major changes in processes and materials that are within the agreed responsibility of Subcon and applicable to the product, prior to implementation. A list of notifiable changes and a price for this extra service are available on request.

9 Production quantity variation from schedule

Subject to availability of components and printed boards, including free issue items, we can accept variations in your schedule of forward quantity requirements up to a maximum of +10% over the top weekly delivery rate specified for the total quantity ordered, i.e. 2200, 2750 and 3300 maxima.

The lengths of forward notice required are:

For +1% to +3%, 4 weeks; for −1% to −3%, 1 week
For + >3% to +6%, 6 weeks; for >3% to −6%, 4 weeks
For ± >6% to ±10%, 8 weeks

10 Validity of quotation

This quotation is valid for 30 days from the date given above and subject to Subcon Ltd's Terms and Conditions of Sale.

11 Value added tax (VAT)

The prices given in this quotation do not include VAT which may be payable in addition, depending on the services given by Subcon Ltd and the location of the country from which the order originated.

12 Payment

12.1 Payment terms for product shipped to customers: 30 days net from Date of Invoice.

12.2 Payment terms for part design, part tooling (including test hardware and software) or other tooling charges: 50% with Order, the remainder within 30 days of dispatch of first prototype or the first production batch, as applicable.

14

Choosing a customer: the subcontractor's viewpoint

This chapter looks at customers from the subcontractor's point of view and highlights one of the most difficult fields in which the executive of a professional subcontract organization has to make judgements. Selecting the right work from the right customers is the task, and by definition this means refusing the wrong work from the right customers as well as turning down the right work from the wrong ones, however tempting the bait may be!

In the early stage of a subcontract company's growth, there is always a tendency to try to be 'all things to all men' and to grasp at the 'big time' jobs too soon. If the customer's project fails, the subcontractor's expenditure in helping to launch the product is completely wasted time and usually profitless effort. The results of the work cannot be offered to any other customer. It is lost money for both parties.

As indicated in the Introduction, the subcontractor is effectively investing in the customer's business and should therefore consider many of the aspects that a bank would look at were it asked to lend money. A corollary of this is that the subcontractor should consider whether the customer is seeing the business between them as a partnership rather than as a master–slave relationship.

Financial status and design track record are the key factors, initially in the decision to quote, then in negotiating the basis for supply and accepting an order.

14.1 Financial and market aspects

The first subject area is the product, its market and the finance required and available. Questions that need to be asked include:

- Is the customer's product viable for its intended market, and who is the end-customer?

It may prove difficult for the subcontractor to make an informed judgement, but every effort should be made to evaluate the product and its potential market within the short time available, even if it means a few phone calls to colleagues in the business.

- Will the market accept the customer–subcontractor team as a valid and financially sound source for the product?

Here again, judgement may have to be made without all the facts. If it is practicable to talk to a few end-customers, this must be done with the agreement of the potential customer and without betraying commercially confidential information—often very difficult.

- Are there any product liability risks outside the subcontractor's and the customer's insurance cover?

One of the key questions is whether it is the intention of the customer or the end-customer(s) to introduce the product in a country where the financial liabilities in the event of mishap could be very high, e.g. the USA. Many European companies are reluctant to pay the high premiums charged for product liability cover in that country.

- Can and will the customer pay bills promptly?

This is where the approach to organizations like Dunn & Bradstreet might provide a clue. The trouble is that very often the available financial information is out of date or, in the case of a subsidiary company, just not available.

- Are the right people, processes and equipment available in the subcontractor's team to make the product to the right standard and the right time-scale?

This is an introspective question, but a very important one, because any significant overstretching of the subcontractor's resources could prejudice the project and harm both parties. There are many case-histories in which the subcontractor executive's failure to ask this clutch of questions has led to difficulties.

- Is the customer's product adequately specified, and can the subcontractor's team be motivated to make it?

Technical assessment is needed to answer the first question. If the product is new, a realistic answer may not be possible. An approach to this problem is given later in the chapter.

The question of motivation should not be ignored, as there have been several cases in which staff have been reluctant to work knowingly on some types of project, for example military weapons, and those in which there is current 'green' lobby activity.

- Will product pricing enable the subcontractor to have an acceptable return?

'Added value per unit of scarce resource time' is more important than gross margin and is the correct yardstick for a subcontractor having a well-mechanized facility.

- Are the cash and borrowing resources of both parties adequate for product launch—especially if the subcontractor is financing the procurement of components and there are major project delays?

In most cases the components constitute the bulk of the production prime cost, and, typically, commitment to purchase begins several months before shipment and often six months before payment.

The likelihood of project delays is a subject requiring sensitive questions and technical assessment of the competence of key members of the customer's staff. These are also dealt with later in this chapter.

In practice, many of these matters are never as cut and dried as the above questions would suggest. Real-life scenarios include a host of unknowns which neither party may feel able to clarify. In any major project, both the subcontractor and the customer will see some similar and some disparate sets of risks. The test of unity of purpose is whether they are able to agree on their order of priority.

14.2 Technical aspects

There is a need for deliberate overlap in some questions so that more than one input can be received from the customer on some of the key topics.

Subjects that should be covered by technical staff in their discussions with the customer's engineers before finalizing the quote include:

- What is the field of application of the product?
- Who is the end-customer?
- Why is surface mount technology needed?

This last is a loaded question because, if the answer shows that it is not vital to the project, a specialist surface mount subcontractor can be sure that margins will be squeezed very hard!

- Is there a possibility that malfunction of the equipment containing the circuit/assembly could bring the risk of harm to people or property?

This is the product liability scene again. It is a fact of life that engineers tend to be more open in their answers than people on the sales side.

- How long is the part to be assembled by the subcontractor intended to last?

This is an essential question when assessing whether the planned electrical and mechanical design intentions of the customer are likely to be 'fit for purpose'.

- What are the relative priorities in the design envelope, for example:

  — Size and shape: are these fixed or can they be varied to reduce cost?
  — What is the cost/price target area?
  — Reliability: is there target MTBF?
  — Does the product have to meet a mandatory user, national or international specification? Is this absolutely immovable, or can it be varied to reduce costs?

These questions are most relevant when the subcontractor is quoting for board layout design as well as assembly.

14.3 Status of the customer's structural design

In asking the following questions, the subcontractor is trying to assess the calibre of the customer's design team and to look for organizational weaknesses that might bring project delay or affect the overall success of the product in the market-place. For example, failure to consider the short- and long-term effects of the method of mounting the board into the customer's equipment can be very destructive. Obviously there will be a desire to correct any design defects before they show up in the field.

- What are the specified operating and storage temperature ranges?
- What proportions of the designed life of the product are in storage and operation?
- During its life, is the product expected to survive regular cyclic thermo-mechanical stresses, e.g. through temperature cycling? For example, is there any regular storage cycle such as there would be for electronics in a motor vehicle left out of doors? Is the circuit within an equipment that is switched on for the day and off at night to cause cycling between +20 °C and +70 °C?
- Are there any acceleration, shock, bump or vibration requirements?
- Is there a full environmental and mechanical specification for the final equipment, and does this require operation in corrosive or otherwise hazardous atmospheres?
- If the board has already been designed,

  — Is there 100 per cent nodal access to all components so that in-circuit test is practicable? If not, why not?
  — How is the board to be assembled into the equipment—for example, by sliding in rack grooves and being pushed into a backplane socket, or by fixing to an aluminium chassis by screws without allowing for differential expansion/contraction?

14.4 Status of the customer's electronic design

The following questions are aimed at determining the status of the electronic design as opposed to the board layout and overall assembly:

- Does the electronic design have the feel of a 'back of an envelope' status?
- What assembly technology was used for the breadboard circuits?
  - How many models were made in that form?
  - Were any of the breadboard models made using surface mount technology?
  - If so, how many and what problems arose, e.g. noise, crosstalk?
- Has the circuit been evaluated for its performance against specification under worst-case conditions? If so, what were the worst-case parameters used; e.g.,
  - Extreme operating temperature range?
  - Supply voltage maximum and minimum?
  - Parametric change in key component values?
- Were the worst-case parameters applied in combination? If so, was this on an absolute or a statistical distribution basis?
- Has there been a sensitivity analysis in respect of parametric values, including predictable drift?
- Can unspecified semiconductor parameters affect circuit performance?

This last point has been the downfall of many projects where the level of functional performance has been squeezed beyond reasonable limits, for example in low power consumption and sensitive high-gain circuitry.

The above technical questions may seem very probing to the customer—even insolent, perhaps! However, they are intended to protect the interests of both parties and should be seen as part of the subcontractor's 'duty of care'.

14.5 Quality aspects

To maximize the company's contribution to the project, the good subcontractor's quality manager will want to know as much as possible about the final product. The following questions are intended to probe the quality aspects of the project:

- 'What visual standards are to be applied to the assembly?'

It can become impractical for a subcontractor to apply different visual standards for all a customer's different products on the same assembly line. The customer will be expected to accept the subcontractor's own common standards and these will have been tailored to 'cover the waterfront'.

However, in exceptional circumstances good subcontractors may accept what they see to be an improvement and incorporate it into their own manuals.

- Are there any special in-process tests called up, e.g. on part-assembled boards prior to adding further components?
- Is 100 per cent 'burn-in' required? If so, what are the thermal and electrical conditions?
- What routine mechanical and electrical tests (if any) are expected of the subcontractor on each batch prior to shipment?
- What tests will the customer carry out at his goods inwards and beyond?
- What will be the basis of the customer's acceptance/rejection of the product? (Refer to Section 18.4.)
- Is there a specific overall quality plan that includes a type approval exercise as well as routine batch testing? If not, who will carry out tests to validate the structural integrity of the design?
- Are any national or international test protocols to be applied, for example as prerequisites to formal release procedures?

14.6 Design for manufacture and test

The following questions are aimed at deciding, if the customer has already completed the design, whether it is suitable for manufacture at the required rate and for the target price. They are very far from being exhaustive, but indicate the method and approach. The subject has been dealt with more fully in Chapter 10.

- How many different component types are used, and what are their dimensions?
- Has the suitability of all components been confirmed for the designed process sequence and soldering conditions?
- Has the board layout been vetted for ease of assembly, visual inspection and rework with the tools available on the subcontractor's assembly line?
- Will the board fit all the equipment to be used in assembly?
- Has the layout included an in-circuit test point for every node in the circuit? Has the layout been arranged so that all such nodes have been brought to one side of the board for ease of testing and avoidance of a clamshell fixture?
- Are all test points at least 1.0 mm in diameter and spaced at 2.5 mm or more? (There are techniques for being able to reduce them to 0.5 mm diameter when reflow soldering is used.)

15

Assessing and auditing a subcontractor's assembly and test activity

15.1 Control of materials

The depth of assessment will depend on whether the subcontractor is providing a complete or partial procurement service or using solely free-issue items. The following text assumes the former.

The task of the subcontractor in keeping tight control of components, boards and other direct materials is much harder than for a typical OEM. The principal reasons for this are the continual flux that comes from a high rate of introduction of new products, the multiplicity of different ownerships, and the management of changes in a profusion of customer designs.

PURCHASING
Starting with the purchasing function, from the organizational viewpoint some relevant questions are:

- Does the buyer attend production and quality meetings regularly?
- How does each buyer receive routine information on post-soldering assembly yields? Can they give examples?
- Are buyers allowed to choose suppliers? If so, under what circumstances and within what limitations?
- What steps are taken to assess vendors?
- Do all purchase orders or the Conditions of Purchase specifically mention good solderability as a requirement or a 'fitness for purpose' criterion in the supply of components and printed boards?

Normally components are purchased against their manufacturer's own specifications, but if reflow-soldering is employed for the product in question, additional requirements should be built into the order to ensure high solderability and therefore minimum rework, the latter being vital to the achievement of good reliability (cf. Section 7.1).

Printed boards and other direct materials do not present the same variety of problems, but still need the technical disciplines demanded by surface mount technology. However, the purchasing specification for printed boards is an important determinant of the approach to both quality and cost. This should cover the items given in Section 8.1.

GOODS INWARDS AND STORES

The strict disciplines required for protecting surface mounted components and boards apply equally in the goods inwards and stores areas. Apart from matters of good housekeeping—typically, labelling to indicate type number, status, ownership or job number—the conditions of storage are more important than hitherto.

When components are supplied in reels, tubes and trays, information on their status should include date of manufacture and date of loading into their reels, tubes or trays. Withdrawal sequences should be based on this information rather than on simple FIFO.

In addition, these items should be stored in conditions that minimize the growth of oxides on solderable surfaces. Humidity is the main enemy, and unless all parts are held in sealed bags or boxes containing dry desiccant, maintaining a steady temperature and avoiding the use of hygroscopic packing materials, e.g. cardboard, are important. Stores in which the heating is switched off at night or over the weekend and on again when the workforce arrive on Monday morning constitute one well-known source of condensation that can occur within the cavities of normal tape reels.

It should be expected that materials such as adhesives and solder pastes would be purchased regularly in small 'day's work' packages rather than in bulk supply. This is because, once a container is opened to atmosphere, deterioration occurs. Also, both of the above can deteriorate in storage, and although solder paste could be usable, its reduced efficiency causes the rework level to be raised above the minimum achievable.

Unless sealed in truly hermetic inorganic containers, solder pastes suffer progressive oxidation of the mainly spherical metal particle surfaces. This can lead to dry joints and solder ball formation during soldering. Some manufacturers recommend storage in a refrigerator, which helps to slow down the rate at which a still sealed solder paste may deteriorate.

Settling out (separation) of the solids and liquids is another cause of long-term degradation, particularly in syringes that cannot be accessed for stirring.

Most adhesives undergo chemical change at room temperature which slowly reduces their effectiveness. Again, storage in a refrigerator may delay the mechanism and improve useful shelf life. However, poor adhesives are not generally regarded as a potential circuit reliability hazard in the same way that solder pastes can be. If a component falls off the board during wave-soldering as a result of inadequate adhesion, it will be replaced by another one at a later stage, and, provided this is done with the right rework tool, there is no reliability problem.

When adhesives are applied on circuits having fine copper tracks and gaps below 0.2 mm (8 mil) wide, the purchasing specification should call for very low free-ion content adhesives. Chlorine and sodium are of particular concern, as in humid atmospheres they can combine with water to form harmful acids which can attack copper and form conducting paths between copper tracks at different electrical potentials.

In all cases, as for boards and components, strict 'EDI–EDO' (earliest date code in–earliest date code out) procedures should be in operation in the stores for adhesives and solder pastes.

To minimize the risk of 'popcorn effects', the procedures used to control unpacking/repacking and logistics between stores and assembly line for integrated circuits require strict adherence to manufacturer's instructions.

Note: Beware managers who say there is no problem because they bake all the integrated circuits that are from opened packages before the devices are soldered. They are ignoring the oxidation caused by baking that reduces solderability, increases rework and can therefore reduce reliability.

Some questions relevant to stores operation are:

- How are the separations arranged, first between the subcontractor's general stock items and goods purchased for specific jobs, and second between free issues belonging to individual customers?
- What happens when the job specifies a mix of the above options?
- Under what circumstances is the chief storekeeper allowed to borrow from one customer's or a specific job's stock to keep another project going? (Of course, the answer should be 'Never'.)
- How is the heating in the goods inwards and stores areas controlled, and is it switched off at weekends?
- If yes, are all components stored in plastic bags containing dry desiccant sachets?

KITTING AND STOCK CONTROL

Kitting costs are highly product-sensitive, and professional subcontractors, although carrying out the operation within the stores area, should regard it as an individual product production cost rather than part of the overhead. The workforce members performing the tasks should be direct employees.

ASSESSING AND AUDITING A SUBCONTRACTOR

This protects those customers for whom the kitting task is not a time-consuming one against carrying an unfair share of the overall costs. For the same reason, one would expect to see kitting as separate action on the formal cost estimate sheet.

Kitting should take place in the stores rather than on the assembly line. For the subcontractor having a wide range of products on the line simultaneously, this can present severe production control problems which must be dealt with, frequently on an hour-by-hour basis. At the end of each week (or day), there will be a review of the availability of parts for the next few days' production and the kitting is planned accordingly. Typical questions to be raised are:

- When the subcontractor is procuring all the components, what happens to part-used reels, tubes or trays? This point is less critical than when they are owned and free-issued by the customer; even so, the manner of their disposal and subsequent storage will give a good indication of materials control activity.
- What happens when components are needed for rework operations? Are they taken at random from those that happen to be around on the shop-floor at the time? Or are they requisitioned separately from stores against a specific job number?
- What paperwork exists in respect of unused components or boards returned to stores? Are the returned items related to job numbers?
- If there is an intermediary stores cupboard on the shop-floor, who has access to it, and how are its contents acquired, kept separated, issued and monitored? Is it kept locked?
- Is production control updated daily/weekly or not at all on the quantities of all items held in stores?
- If not, how is purchasing alerted to the likelihood of stock exhaust due to poor yields or a major batch loss, so that replacement components and boards can be obtained before production of a circuit is stopped?

15.2 Process control

THE PRODUCT FILE

The first item to check is the content of the product file. This is a group of documents that is passed to the production department—usually from a design group or a customer liaison engineer. Its purpose is to define the individual product and its exact method of manufacture and test.

The file should, at a minimum, contain the following data:

- A description of the product function and, whenever practicable, that of the final product of which it is a part. As previously indicated, the latter can be very helpful in motivating the workforce.

- A complete parts list.
- An electrical circuit diagram.
- Bareboard track/gap layout diagrams for both sides of the board.
- A printout of the complete assembly and, if appropriate, each sub-assembly. This should show the position and value of each component on each face of the board(s).
- A complete process sequence chart, with any necessary limitations or warnings, e.g. advice needed to protect the board, components, materials and (in some cases) the production equipment used.
- A list of components that are not checked by the in-circuit test rig. These will require particular attention in visual inspection. Refer to comments on in-circuit test percentages in Chapter 18.
- Packing and shipping requirements.
- Batch size.
- The target process times, machine times and stage yields used in the cost estimate.
- Whom to call in the company in case of queries. Also, if appropriate, the names and phone numbers of the relevant customer contacts.
- A copy of the electrical test specification and acceptance limits.
- The relevant visual inspection standards.

PROCESS CONTROL METHODS

The second assessment item should be the processes and the methods applied in controlling them. Taking them individually, the auditor would expect to see the following.

1 After solder paste deposition

Regular sample checks looking for misalignment of paste and pads, excess paste, insufficient/missing paste, spread/slump and smudging. In high-volume production, there would be hourly measurement of mean wet paste thickness using a non-contact method.

The cleaning of printer screens and the control of paste are instrumental in minimizing the need for rework. The repeated mixing of old paste (i.e. where the container was opened more than eight hours previously) with new paste should be avoided as this leads to poor soldering and the formation of solder balls. Print screens that are not completely free of old paste can bring the same problems when the new paste is pushed to and fro and absorbs the old. This applies to the outermost corners of the screen or stencil as well as to the aperture areas.

Other methods of applying solder paste include automatic and manual dispensing from a syringe. In all instances, maintaining a steady machine and paste temperature is critical to controlling the amount deposited. Prin-

ASSESSING AND AUDITING A SUBCONTRACTOR 163

ters sited near hot ovens or beside sunny windows are likely to experience control problems in summer.

2 After adhesive placement
One would expect to see either automatic 100 per cent or regular sample checks looking for misplacement of adhesive, excess and insufficient/missing adhesive, spread/slump and smudging.

Methods of deposition include syringe dispensing, screen or stencil printing, pin transfer printing and secondary cylinder dispensing. In high-volume production using a syringe integral with the automatic pick-and-place machine, there would be on-line TV monitoring of the area wetted by each adhesive dot, with automatic feedback and control of dispensed quantity.

Deposited adhesive dot quantity and contour is even more sensitive to temperature variation than solder paste, with the added complication that, if the spread is excessive, the dot height may not be sufficient to make contact with the underside of the component to be attached. The component will then fall off when the board is inverted for insertion or wave-soldering.

Close bi-directional control of the dispensing mechanism temperature is essential to all methods of deposition. Incorrect adhesive location and quantity on the printed board result in excessive rework and consequent reliability risks.

3 After component placement
Inspection for misalignment, missing, reversed or wrong components can be monitored automatically by a scanning machine. Looking for damage caused by the operation to either the board or the components requires manual intervention.

4 After reflow-soldering
In addition to the faults looked for after component placement, the post-soldering inspection should include checks on all joints for excess solder (including bridging), insufficient solder, solder balls and splashes, 'tombstones', 'crocodiles' and other forms of open circuit and any thermal or mechanical damage suffered by the board itself.

5 After wave-soldering
The basic inspection is similar to that used for reflow-soldering, but there are minor differences in that solder icicles would be added under the 'excess solder' category and 'skipped' joints are included under 'insufficient solder'.

6 After cleaning (if applicable)

The board should be examined visually without magnification for residues, tide-marks and any solder balls or splashes that were not removed by the cleaning process itself.

Where the cleanliness requirements are more stringent, typically for circuits used in adverse environmental conditions or when they are inaccessible for repair, either the quantification of residual contaminants using chemicals or the measurement of inter-track surface resistance is applied. The two methods most commonly used are the 'Ionograph' test and the surface insulation resistance (SIR) test. Since the SIR test has to be preceded by several days' exposure to damp heat, the former is more often seen as an on-line monitor for rapid use on a sampling basis.

7 After in-circuit electrical test

The main component defect categories will be short-circuit, open-circuit and parametric. In addition, the test will be expected to pick up faults such as missing or reversed and wrong components, which may have been overlooked in previous inspections.

15.3 The quality function

The structure, reporting chain and activities of the quality function can vary widely between subcontractor companies. Any organization in which the quality manager does not report directly to the chief executive or managing director must be viewed with caution. To this end, the auditor will expect to investigate the following points.

THE ORGANIZATION CHART AND STATUS OF THE PERSON IN CHARGE OF THE QUALITY FUNCTION

Ideally, this person should be a member of the board of directors. In any case, he or she should not report via the manufacturing or technical/design/engineering functions. The reason for excluding the latter routes is that so many of the quality problems seen on surface mount assembly lines are a direct result of inadequate board design.

THE AUDIT FUNCTIONS CARRIED OUT BY THE QUALITY FUNCTION

These should include all departments. Often the sales and marketing activities are ignored as having no bearing on quality matters. Of course this is a wholly mistaken idea when dealing with subcontracting activities, especially where sales office staff form the main link with the customer or when product safety is an issue.

THE TRAINING PROGRAMMES

In companies that do not carry a separate personnel department, the training programmes for all workforce members are often managed by the quality function. In any event, they should definitely be included in the internal audit sequence. This interest should not be limited to quality matters.

In surface mounting, the shop-floor training periods are several times as long as for through-hole assembly and are especially important for inspection staff and for operators who are expected to check their own work. It is normal that those with the best combination of intellectual and manual skills will be chosen to carry out rework operations and later to be trained as inspectors.

ROUTINE MONITORING

The monitoring items listed below against various departments are not exhaustive and omit many of the normal requirements. They are intended only to highlight areas of particular importance in the application of surface mount technology.

The monitoring of technical/design procedures should include:

- The operation of the Change Note system
- The preparation and content of the product dossier
- The application of design rules and checking of layouts prior to manufacture

The monitoring of the purchasing function activities should include assessment of:

- The control of product safety information
- The steps taken to purchase components with good solderability
- Vendor assessment and auditing activity on approved suppliers
- Positive action to get feedback from the assembly line to activate timely reordering of parts lost through poor yields
- Positive action to identify unsatisfactory suppliers

The monitoring of stores should include:

- Operation of the EDI–EDO system for adhesives, solder pastes, components and printed boards
- Anti-static precautions in handling and kitting
- The protection of solderability
- The segregation of components by project and ownership

The monitoring of assembly activity should include:

- Performance of all assembly activities involving operators

- Control exercised over all processes and machine operations
- Checking of components, joints and boards after assembly
- Control of rework activities and loops
- Control of product status at all stages
- Collection of statistical process control (SPC) data at key stages and its use in corrective action
- Maintenance of agreed traceability routines

The monitoring of training activities should include:

- The upkeep and status of workforce training log books

The monitoring of packing and shipping should include:

- Consistent use of anti-static precautions
- Effective separation of individual circuits to avoid damage in transit

THE CONTENT AND UPDATING PROCEDURES FOR THE QUALITY MANUAL
The new disciplines required in all departments demand a much better level of professionalism than in the past.

The Quality Manual should contain a section for each department defining its role and responsibilities in quality matters. Most subcontractors adopting surface mount technology as an additional capability to their through-hole work have found the need for drastic revision of their previous document.

The dates of updating in the various sections of the manual should be checked against those called for in the internal departmental audit corrective action schedule.

15.4 Visual standards, inspection and rework procedures

Study of the company's surface mount and mixed-technology visual inspection standards is a key aspect of the assessment. Equally, the method of applying them and of controlling resulting rework has a direct bearing on the quality of shipped product.

It should be borne in mind that the visual inspection of surface mount components and their solder joints is a far more taxing task than for conventional through-hole boards. Current wisdom is that it is unwise to expect the required concentration for longer than two hours at a stretch or for more than four hours per day in total. The subcontractor's policy on this issue should be checked.

Of course, this problem is correctly addressed by devising methods of avoiding the need for inspection. The range of options ranges from 'zero defect' manufacture to fully automatic 100 per cent inspection of every joint, both of which are best suited to continuous very high-volume manufacture on single product lines. Only in those circumstances can they give a reasonable payback on the heavy capital investment and high data collection and feedback costs incurred.

Inspection methods can be categorized as visual, thermal, mechanical, acoustic and thermo-mechanical, all except the visual method being optionally destructive (D) or non-destructive (ND). Examples of each are given in the following lists.

Manual visual techniques requiring human decisions (ND)
- Illuminated magnifiers
- Binocular microscopes ± mirrors
- Stereo projectors
- CCTV ± mirrors
- X-ray
- Laser doppler vibrometry

Automatic visual techniques (ND)
- X-ray laminography
- Multiple image processing
- Laser profile scanning

Visual accept/reject criteria are based on a comparison of wetted areas, contours and meniscus shapes with visual standards or, for automatic techniques, with library data based upon them.

Thermal technique: laser heating/infra-red scanning (ND)
Accept/reject criteria are based on comparison of rise/decay times for defined thermal inputs to the joint compared with library data.

Mechanical techniques
- shear strength via push or torsion (D or ND)
- Pull strength (D)
- Vibration (D)
- Shock (D)
- Acceleration (D or ND)
- Substrate bending (D)

Accept/reject criteria are based on shear and tensile strength test results, or survival rates in repeated mechanical deformation of populated substrates.

Acoustic technique: pulsed sonar beam (ND)
Accept/reject criteria are based on the reflected response to sonar beams and comparison with library data.

Thermo-mechanical technique: temperature cycling (D or ND)
Accept/reject criteria are based on survival rates in temperature cycling.

In all of the automatic techniques, the cost of analysing and relating outputs to standards defining acceptability and rejection and the cost of setting up for individual circuit requirements can be high. In addition to identifying the faults, their location and rapid means of rework may also require expensive equipment.

For the majority of medium- and small-sized subcontractors, the production quantity range covered per circuit means that the choice lies between two main courses of action.

1 Visual inspection of all solder joints before electrical test, using manual methods
This is the most primitive of all approaches, and its adoption is as likely to stem from poor layout design or poor-quality components as from incorrect processes. It is truly the worst of all worlds, but subcontractors are sometimes lumbered with their customers' design mistakes and must do the best they can in the circumstances.

2 Visual inspection after electrical test
For electrically good circuits, this will be on a sampling basis; e.g., 100 per cent of the joints are inspected on 10 per cent of the boards, but for the failures all joints on all boards will be checked. The criterion here is where the defect level is low enough to allow the cost of testing without prior inspection to be justified financially. Obviously, at a given defect level in parts per million the critical point is more likely to be reached with circuits having a low pincount than with large boards containing several hundreds—or even thousands—of solder joints.

Because the advent of surface mounting has greatly increased the number of solder joints on a typical board, the likelihood of rework has risen to the point where, for large boards, two or more rework/inspection cycles may be required and should be both planned and costed (see Table 19.2).

The cost of these manual operations makes it essential for management to know the extent to which such cycles are occurring for each product. A corollary is that, to relate cost and reliability to the rework function, all boards should be marked each time they fail a test as well as when they have passed it.

ASSESSING AND AUDITING A SUBCONTRACTOR 169

A second, but not secondary, issue is to note who is actually doing the rework. The most economic way for a subcontractor to tackle this is to ensure that whoever does the visual inspection also does the rework. The reason is the excessive time required for the alternative; for example:

- The inspector picks up a small adhesive marker arrow from a sheet.
- It is attached to the board, pointing exactly to the defect.
- The board is passed to an operator in the rework section.
- The rework operator identifies the fault and corrects it.
- The adhesive arrow is re-attached—still pointing to the defect area.
- The board is passed back to the inspector.
- The reworked part is re-inspected.

In this work sequence, not only is the time taken far too long, but there is also a significant risk of data loss or corruption in the transaction.

It is considered quite satisfactory for the checker or inspecting person to do the rework provided there is periodic auditing by a third party. In practice, the first inspector/checker is usually a member of the production staff, while the auditing is done by quality staff. Where the first check and the rework are both carried out by a member of the quality department, a second member should provide the audit function.

Aside from this, the process should be monitored by logging the rework on each circuit using a rework chit. This identifies the batch number and the person doing the rework, categorizes the defect and serves as a means of data transfer and data collection for statistical control purposes. It also helps to pick up regular faults requiring corrective action alongside the use of measle charts.

Data collection for statistical process control and the use of it to gain corrective action is another point to watch for. Examples of typical approaches at two depth levels are given in Figs. 15.1 and 15.2. One would expect to see these applied to a range of key processes, e.g. after depositing solder paste, after component placement, after soldering, after cleaning.

15.5 Electrical testing

A key point to look out for is the availability of on-site in-circuit test probe systems with sufficient nodes to cover the known requirement. The number of circuits awaiting retest following rework operations is a useful guide to efficiency if the total throughput is known, but this can also be an indication of poor design by the subcontractor's existing customer(s).

If the test equipment has a printout system indicating each defect found, this should be in the form of a simple instruction to the rework operator rather than a statement that is understandable only by an electronics engineer, e.g. 'R71 high, replace' as opposed to 'V1/V2 low'.

Process: Reflow Soldering		Sample Size		Machine							
Checking Frequency		Operator		Date							
Board Type/No.											
	Defect Classification Level 1										
1	Misalignment										
2	Missing Component										
3	Reversed Component										
4	Wrong Component										
5	Damaged Component										
6	Component on Edge										
7	Bad Wetting/Dewetting										
8	Excess Solder/Bridging										
9	Insufficient Solder										
10	Solder Balls										
11	Tombstone/Crocodile										
12	Damaged PCB										
13											
14											
15											
Total Defects											
Baseline: No. of Solder Joints											
Parts per million											

Fig. 15.1 Process control chart: defect classification level 1

Process: Reflow Soldering		Sample Size		Machine								
Checking Frequency		Operator		Date								
Board Type/No.												
	Defect Classification Level 2											
1a	Misalignment, General											
1b	Misalignment, Local											
2a	Missing Component											
3a	Reversed Component											
4a	Wrong Component											
5a	Damaged Component											
6a	Component on Edge											
7a	Bad Wetting, Dewetting, General											
7b	Bad Wetting/Dewetting, Local											
8a	Excess Solder, General											
8b	Excess Solder, Local/Bridging											
9a	Insufficient Solder, General											
9b	Insufficient Solder, Local											
10a	Solder Balls General											
10b	Solder Balls Local											
11	Tombstone/Crocodile											
12	Damaged PCB											
13												
	Total Defects											
	Baseline: No. of Solder Joints											
	Parts per million											

Fig. 15.2 Process control chart: defect classification level 2

If the subcontractor claims approval to a national or other specification, check the labels on a few pieces of electrical test gear to see if the calibration date system is working effectively.

15.6 Traceability requirements

Traditionally, traceability requirements have been dependent on the application of the circuits being manufactured. In some military and all life-support situations, the ability to relate every component and board to the original manufacturer's production batch number has been paramount.

Mechanized high-volume assembly techniques have made this degree of perfection in traceability not totally impossible, but very difficult and certainly very costly.

Most discrete surface mount component sources are hard to identify by visual means. They are often unmarked because there is not enough surface available for parametric information, let alone a manufacturer's logo. At the same time, minor differences within the specified standard outline shapes and subtle variations in structure may enable their origin to be detected by experts.

In any event, the suggestion that buyers should insist that the date of manufacture appear on all boards, reels, tubes and trays helps solve traceability problems.

Sequential numbering in the copper by the board manufacturer is practicable at the panel level, but extending it to individual circuits in stepped and repeated arrays within a panel is not. Missing numbers due to production losses, first those of the board manufacturer and subsequently those of the subcontractor, would expose yield data to the customer which could be considered embarrassing.

To be effective, any alternative system involving marking or the attachment of labels must be proof against malpractice and also capable of withstanding the assembly and cleaning processes. One such method is to use a laser or other means of burning or imprinting data indelibly just before the board manufacturer's shipment stage. This approach involves an expensive piece of equipment capable of x, y movement across step-and-repeat panels and an easily programmable control system. Any defectives within the array would not receive a number.

An alternative, less secure and less precise but cheaper, approach is for the subcontracting assembler to label or otherwise mark board numbers at the point of dispatch. The traceability of the board and its contents is then dependent on:

- The subcontractor's ability to relate the board numbers to the assembly batch numbers and the latter to both reel/tube/tray numbers and goods inwards batch numbers

- The component manufacturing firm's ability to connect the reel/tube/tray data to its own production batch information

This latter path can be a long and tortuous chain stretching around half the world's circumference. Things are made especially difficult when components manufactured overseas are purchased in small quantities from local distributors; often they are asked to supply less than the content of one reel or tube and the identities can easily be lost.

In practice, most professional subcontract assemblers are prepared to relate their shipments to goods inwards material identity only when all the components are procured uniquely by the customer firm and where the pick-and-place machine feeder bank is bespoke to that firm. Outside these constraints, again it becomes impracticable—but not impossible—to control events to the required degree in the stores and in highly mechanized shop-floor operations.

At the same time, it must be appreciated that hand assembly may not be a satisfactory solution either, because the reliability of surface mount component assemblies is so closely linked to the degree of control exercised over the soldering process. This cannot easily be ensured by manual methods.

Obviously, it is not much use having good traceability for the parts if there is little or no control over the assembly processes used. On this issue one would expect to see that the process setting for each traceable circuit is listed in its product dossier and that the correct settings are actually adhered to. This is crucial for the soldering operation and for any subsequent cleaning processes involving ultrasonics.

A further question concerns the traceability of manual workmanship to individual assembly operators. Generally this gives fewer problems because it involves either manual work or the setting of machines at specified times and dates. Personal signatures on batch cards, operator time booking sheets or travelling circuit cards can be designed to give the required information, though this is not always done. If the subcontractor is collecting data on the cost of each different circuit made, then workmanship traceability will be achieved without much difficulty because job bookings and related dates can tell the story. This is a useful guideline.

A typical traceability chart is shown in Fig. 15.3.

15.7 Quality assurance testing

At a minimum, the professional subcontractor will have a temperature cycling chamber, preferably two. One should be capable of operating between room temperature and -40 °C and the other between -10 °C and $+70$ °C. For military and aerospace applications, these temperatures need extending to cover the range to -55 °C and $+100$ °C.

Force or torsion gauges of varying maximum force ranges are needed to

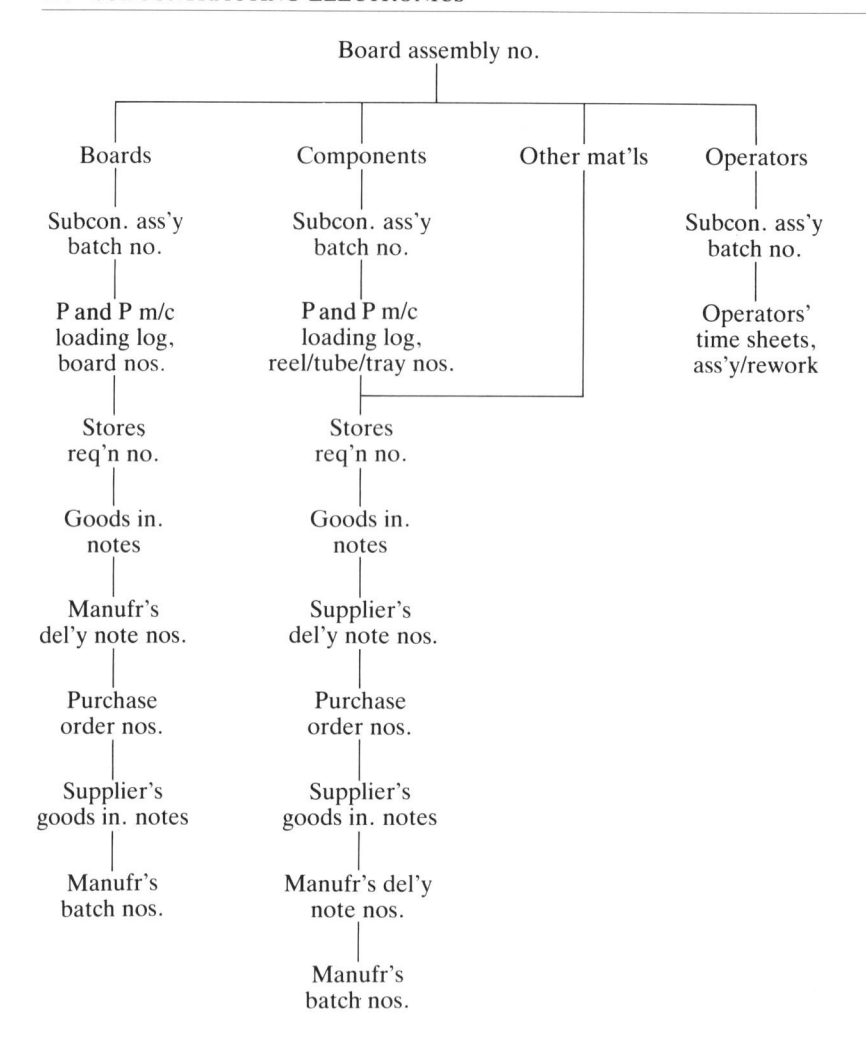

```
                        Board assembly no.
        ┌──────────────┬──────────────┬──────────────┐
      Boards        Components    Other mat'ls    Operators
        │              │                             │
  Subcon. ass'y   Subcon. ass'y                 Subcon. ass'y
    batch no.       batch no.                     batch no.
        │              │                             │
    P and P m/c    P and P m/c                   Operators'
   loading log,   loading log,                   time sheets,
    board nos.   reel/tube/tray nos.             ass'y/rework
        │              │
      Stores         Stores
     req'n no.      req'n no.
        │              │
     Goods in.      Goods in.
      notes          notes
        │              │
     Manufr's       Supplier's
   del'y note nos. del'y note nos.
        │              │
     Purchase       Purchase
    order nos.     order nos.
        │              │
    Supplier's     Supplier's
  goods in. notes goods in. notes
        │              │
     Manufr's      Manufr's del'y
    batch nos.     note nos.
                       │
                    Manufr's
                   batch nos.
```

Fig. 15.3 Typical traceability chart

validate the ultimate adherence strength of adhesives and of surface mounted components. Typically, a complete set of gauges will need to operate from 0 to 0.25 kg, from 0 to 1 kg, from 0 to 5 kg and from 0 to 10 kg.

Burn-in facilities tend to be bespoke to particular projects and would not be regarded as general background quality assurance equipment.

15.8 Anti-static precautions

The degree of rigour applied by a subcontractor in protecting static-sensitive components is usually a measure of two things: first, the awareness and level of understanding in the workforce of the risks to circuitry from static electricity, and second, whether they have experienced a real disaster from failure to take the necessary precautions.

From the technical standpoint, most professional subcontractors take the view that all circuits must be treated as static-sensitive, irrespective of reality. The reason is based not on altruism, but on the fact that, with a wide variety of customer circuits on the assembly shop-floor, it is often impractical to differentiate between those that are sensitive and those that are not.

A factor not always recognized is that semiconductor devices are not the only items requiring protection. In the steady pursuit of more microfarad-volts per cubic centimetre, today's multilayer ceramic chip capacitors are being fabricated with progressively thinner dielectric layers, e.g. with thicknesses in the region of 20–25 microns. The interleaved metallic electrode layers are not self-healing, and voltage spikes can cause permanent short-circuits.

If all static charges were to cause immediate and complete breakdown, the result would almost invariably be readily detectable circuit failure. Serious problems arise when the semiconductors and capacitors suffer damage that is not fatal but creates a 'near miss' situation which is not obvious and is therefore not detected at the time of final testing and shipment. Subsequent operation in the field becomes sufficient to complete the destructive process, and an equipment failure results. The unpredictability of this event sequence, and the often hidden nature of static damage, make strict precautions even more important, but also more difficult to impose.

A suitable analogy is the case of a military establishment seeking protective security arrangements against terrorist attack. Success for the terrorists can be very dependent on human frailties among security staff and their managements, for example failure to provide proper training, lack of imagination, boredom.

The logical approach is to minimize dependence on human error by building in protective barriers and status indicators. This process starts with the electrical design of the circuit and ends with the complete mechanization of all processes, including rework and packing and shipping. This 'anti static nirvana' being unattainable for all but the highest levels of production rate, lower degrees of protection must be devised and accepted.

What should a quality auditor visiting a professional assembly subcontractor expect to see?

Of equipment, facilities and materials
- With respect to the handling of all components and assembled boards, all machines should have static earth cables fitted and at least two sockets for operator's flexible earth cables to plug into, one for the

machine operator and one for an inspector or technician. This requirement should apply at least to pick-and-place machines, to soldering machines, cleaning equipment, test equipment and, of course, to the stores, kitting and packing and shipping areas.
- All benchtops and inter-stage work storage shelves should have conductive work surfaces fitted with static earth cables. Each operator seating position should have two sockets.
- All interstage and work-station storage racks and shelves should be of conductive material or have conductive surfaces, and all should be fitted with anti-static earth cables.
- Flooring should be of conductive material and properly earthed.
- All packing materials should be conductive.

In this context, care should be taken to ensure that the conductivity is provided by impregnated or surface materials which do not degrade the solderability or other properties of the product through ionic contamination.

Of the workforce
- All operators, inspectors, technicians and supervisors should be wearing wrist straps connected to their personal flexible earthing cables at all times when working in the stores, kitting, production, inspection, test, and packing and shipping areas. Where the use of cables is impracticable, the wearing of conductive attachments to shoes should be compulsory, but they require conductive anti-static flooring to be effective.
- When handling a product, for example on a bench or attending a machine, personal cables should always be plugged into anti-static earthing sockets.
- When transporting loose components or assembled boards by hand from one place to another, these should be placed on conductive plastic materials provided for the purpose.

15.9 General housekeeping and discipline

Walking into a subcontractor's surface mount or mixed-technology assembly area, apart from a general impression of tidiness and cleanliness, some key questions to ask are:

- Are the vast majority of all heads down, focusing on the work?
- Do all, but all, the workforce wear white or pastel-shade coats, and are they reasonably clean? Are all visitors asked to wear them too?
- Is smoking prohibited?
- In areas making products for 'hi-rel' applications, are shoe covers being

used? If white headcaps are being used, are workforce members defeating the objective by letting their hair dangle outside them?
- Are the work-station benchtops and storage shelves reasonably clean?
- Are there many pieces of gear and odd items sitting on the floor?
- Are there many small components or short lengths of wire clippings lying on the floor? (They tend to congregate near the pick-and-place machines and under benches where hand operations are in progress.)
- Look at the methods of stacking and transfer of circuits at and between work-stations. Do they effectively prevent touching of adjacent boards and permit easy identification of product type and status, e.g. awaiting inspection, passed inspection?
- How are the solder paste printing screens and stencils stored when not in use? Are they scattered over nearby surfaces, lodged in open shelving, or stacked neatly within closed cupboards as they should be?
- If a batch cleaner is being used and is still operating with a CFC liquid, is the cover in place when no boards are being processed within it?
- Do the assemblers know the destination or, more important, the application of the boards they are working on?

15.10 Trouble-shooting capability

The ability of a subcontractor to spot and diagnose the underlying reasons for the type of defects seen on a surface mount assembly line is an important confidence-building element for the customer. Many of the difficulties demand a knowledge of chemistry and physics well above the level seen in most conventional through-hole assembly facilities.

A microscope with colour camera attachment is the most useful tool for communicating visual fault information to customers. This needs to be a high-definition unit as components and their joints are getting progressively smaller.

Another type of visual aid frequently seen is the twin-position training microscope designed so that trainees or visitors can observe a manual work operation in parallel with the operator.

While one would not expect to see an electron microscope or a surface analysis equipment in the corner of the factory, access to one in a nearby university or laboratory is a useful adjunct which can help solve problems needing spectrographic analysis very quickly, for example checking on unknown residues.

A combined measuring and metallurgical microscope is another useful tool more often seen, typically with a magnification capability up to $\times 100$. Where the longest dimension of printed boards accepted by the subcontractor is large, e.g. greater than 30 cm, the inspection requirements would probably need separate pieces of equipment as the measuring microscope would be unlikely to have a bed with sufficient travel capability.

Also in this category is the special microscope/projector for looking at the internal surfaces of plated-through holes and vias. One would expect to see this type of equipment where the subcontractor is procuring multilayer printed boards in quantity.

16

Rework and repair

[*Note*: For the convenience of readers, this chapter is reproduced in abridged form from Martin Wickham's Chapter 9 in D. Boswell and M. Wickham, *Surface Mount Process Control, Quality and Reliability*. For details, refer to the Introduction.]

16.1 Background

The rework and repair of surface mount component assemblies is certainly more difficult than for conventionally mounted through-hole items, but with care the task can be completed to give satisfactory joints. Most of the techniques and limitations discussed below are applicable to initial hand assembly as well as to rework.

This chapter makes no distinction between customer and subcontractor, because either or both may be involved in the rework and repair of circuits. It is included in the book because the subcontractor's capability and control of rework are vital to the protection of surface mount product reliability.

There are four basic requirements for successful rework: good printed board layout design, selection of the correct rework equipment, sufficient manual skill, and adequate training.

The formation of intermetallic compounds during remelt means that touching up some joints may do more harm than good, particularly in the case of components with sensitive termination materials. Because of these changes in micro-structure, joint fatigue life—as measured by the number of temperature cycles endured before fracture occurs—can be significantly reduced.

For companies used to conventional through-hole assembly, effective control of surface mount rework quality is likely to require the extension of procedures and new disciplines for identifying product status. Their degree depends very much on the total board pin (joint) count, as this strongly influences the likely number of rework, cleaning, inspection and test cycles through which each assembled board may pass.

16.2 Assessing rework equipment and processes

If rework has to be done, then ideally users would prefer that one tool or equipment covers all requirements. Unfortunately, rework tooling and techniques are strictly matters of 'horses for courses'. When choosing a method for a particular component, it is convenient to assess it against a set of ideal requirements. These are:

- No inherent operator health or safety hazards
- No damage to the reworked component
- No damage to adjacent components
- No damage to the printed board
- Built-in preheat to reduce thermal shock
- Simple operation—minimum skill required
- Single tool for component removal and replacement, solder addition and removal
- Minimum time to complete rework operation
- Small tool size to permit high component packing density
- Vacuum pick-up to remove component after reflow
- x, y, z and theta movements of board and/or head
- Placement accuracy assistance for fine lead pitch devices
- Integral magnified visual assistance
- Closed-loop temperature control
- Portability for field servicing (if required)

Clearly, no single equipment exists which meets all of these points. Assemblers may wish to give more weighting to some requirements than others, depending on the application of the product and the salvage priorities; e.g.,

- Save the main printed board assembly at all costs.
- Save the component because of its high cost or non-availability of a replacement.
- Save both board and component for re-use or analysis.

To maintain quality and reliability in reworked joints, operators need above-average manual dexterity and good eyesight (particularly good acuity), and they should not be colour-blind.

In many companies the most efficient way to handle rework is to have it done by the staff who carry out the visual inspection. This is because in touching up joints it can take almost as long to mark the fault with a tiny adhesive-backed plastic arrow as it does to do the rework. If we then add the extra time and control procedures involved in passing work to and fro between rework and inspection stations, the savings are obvious.

However, this approach requires careful monitoring. Inspectors should be trained to know when to do the rework and when to blow the whistle on production and make them correct repetitive faults.

16.3 Key problems in surface mount rework

(A) NON-MARKING OF COMPONENTS AND SUBSTRATES

With lack of marking on many components and the tendency to omit 'ident' or 'legend' on surface mount printed boards, the task of rework requires rigid disciplines as well as extra care. A full component layout diagram should be supplied to each rework operator and/or inspector, together with a detailed components list.

Wherever possible, replacement components should be extracted from their protective tapes or tubes as they are used. Decanting them into trays is not recommended unless the latter are covered and kept scrupulously clean, and even then they do not benefit from the collective jostling involved.

If excess quantities are loaded into the trays and lie around in the open, they become exposed to moisture, factory dust and contamination, including human detritus and coughs and sneezes. These bring significant risk to solderability and to reworked joint quality.

To prevent confusion, any surplus or loose components without marking on their bodies should be carefully identified as to value, type and batch number and stored in a protected space near the workplace. Use vials rather than polythene or polypropylene bags.

Where printed legend on the board is completely omitted, as is recommended by myself and others, a component co-ordinate grid system is often used to identify their respective positions. To assure correct replacement, rework operators should be trained to note the polarity of all defective diodes, electrolytic capacitors and integrated circuit packages before removing them—even when incorrect polarity is the reason for their action.

No attempt should be made to replace a defective component with one of the same value but having a different package footprint size. Such a component may nearly fit on the same pads, but may not make a reliable, long-life connection.

Never fit conventional wire or tape-leaded component leads to pads intended for surface mount terminations. If this is absolutely unavoidable, ensure that the component body is firmly glued to the board before soldering.

(B) PREHEATING COMPONENTS

Where it is known that the rework equipment to be used for replacing components is capable of imparting severe thermal shocks, e.g. in the case of a soldering iron, it is important to preheat the components. Usually this is done in a small oven adjacent to the rework station which is set to the maximum storage temperature of the components.

(C) HANDLING OF COMPONENTS

The main tools for handling individual components for manual placement on printed boards are the vacuum pencil and tweezers. If the latter are employed for handling chip ceramic components, use of conductive plastic tweezers is strongly recommended to minimize the risk of damage to their brittle ceramic bodies. With these tools, to minimize contamination risks components should be handled only via their bodies, not by their terminations.

With vacuum pencils, care must be taken to keep the small suction pads scrupulously clean and to replace them frequently. Through constant use, their pick-up surface becomes coated and later impregnated with a thin layer of dirt and grease, and chip solderability can be impaired through contact with them.

(D) RE-USE OF REMOVED COMPONENTS

Most component manufacturers are unwilling to give effect to their guarantees if their product has been removed from a printed board and remounted.

While there is always a risk of damage arising, it is possible, if need be, to perform the removal and re-use operation successfully. Whether, as a result, the 'patient' suffers early death must be at the risk of the person authorizing the work. However, it is sensible to assume that some reduction in reliability will occur.

Whichever method is applied, clearly some components are more at risk than others, and the choice of tool and skill of the operator are both critical. Devices that are usually more sensitive to removal and re-use are:

- Multilayer ceramic chip capacitors
- LEDs
- ASICs in PLCC or quadpack format
- Wave-soldered precision resistors
- Large SOICs (> 16 leads)
- Wave-soldered quadpacks
- Any component for which the data sheet specifically disbars re-use

Automatic rework machines with sophisticated control of times and temperatures are preferred to manual methods on reliability grounds.

(E) PRINTED BOARD LAYOUT DESIGN AND SPACE CONSTRAINTS

Many users adopt surface mount technology because of its potential for cost-effective miniaturization. However, the printed board layout designer should avoid the use of 'technology for technology's sake', and a careful compromise must be sought between the conflicting requirements of reducing 'real estate' and ease of assembly, electrical test and rework.

Often this means resisting the pressures from enthusiastic project managers who may advocate squeezing out every last drop of performance and achieving the ultimate in small size—all without sufficient regard to product manufacturing cost, testability and reliability.

Our own experience in visiting European companies under the auspices of the Surface Mount Club indicates that project managers all too frequently fall into this trap and then blame the technology when things go wrong on the production line. The quality department must seek a positive role in this arena by contributing feedback data in support of intelligent compromise before the event, not after it.

If components are too close, adjacent or replacement components can easily be damaged during rework. Nearby solder may be reflowed a second time, leading to reduced mechanical attachment strength and the risk of dry joints. For those components that have been glued and wave-soldered, sufficient clearance should be allowed around the devices so that they can be twisted through 90° in one direction to shear the adhesive while all the joints are molten.

Successful removal of large multi-lead integrated circuit packages involves the use of hot gas or heated electrode tools. Sufficient clearance around the package to permit the rework head to surround the device completely is essential. Refer to the third case history in Chapter 20.

(F) HEAT SINKING EFFECTS

Where large ground planes or heat sinks are present in a printed board substrate, these will conduct heat away from the component being reworked. Extra heat, perhaps for longer periods, is then required, which in turn can lead to damage to components or the board. The fact that the solder joints may not reach reflow temperature is no guarantee that the component or the board have not been overheated.

This is a design problem which must be tackled at the printed board layout stage. Wherever possible, any component termination that may need rework, including leaded through-hole types, should be thermally isolated from any ground plane or integral heat sink by a short length of copper track.

A further exposure to heat can come from use of a much larger soldering iron bit. If such a crude implement has to be used, the likelihood of its touching and damaging adjacent components can be high.

Where a heat sink has to be attached to a component, either it should be of a type that is removable without disturbing or stressing the solder joints, or, if not removable, it must not impede access for the appropriate rework tool and should not act as a significant sink for the heat applied by the rework tool.

Alternatively, it may sometimes be necessary to protect a component body from excess rework temperature, e.g. by clipping a local heat sink between the body and the solder joint. A specially formed crocodile clip can fulfil this function.

(G) CHOICE OF ADHESIVES FOR WAVE-SOLDERING

Here again there are conflicting requirements. For assembly, the adhesive needs to have sufficient strength at soldering temperatures to ensure that components are not dislodged by the wave. In contrast, for rework a lower (ideally zero) strength is required to allow easy removal of the part from the board.

(H) PRINTED BOARD MATERIAL TYPE

To ensure minimum damage to the printed board during rework, the base laminate should be a good-quality epoxy–glass type with a high copper peel strength.

Where high packing density is required, the use of inferior laminates or copper-clad paper-phenolic boards can easily lead to problems with pads peeling away during rework. This may result either in the scrapping of complete assemblies or in expensive repair of damaged copper areas.

(I) COPPER PAD AND TRACK LAYOUT

If space on a board is at a premium or signal paths must be kept very short, designers will often route a track between adjacent device pads spaced at 1.27 mm (0.050 in) pitch. In such cases, tracks should be covered with a wet film photo-imageable resist to minimize the risk of lifting during rework operations.

Tracks between pads at 1.0 mm (0.040 in) pitch and below are not recommended because of the high chance of damage during rework.

16.4 Selection of suitable rework equipment

When reworking conventional through-hole joints, a vacuum desoldering tool will enable most common device types to be dealt with. Unfortunately, with surface mount technology there is no single equipment that will cover all requirements efficiently. Users will find that they need as many as three or even five different tools as well as a variety of different heads for each. The selection of rework methods and tools depends on a number of factors.

(A) DEPENDENCE ON COMPONENT TYPES MOUNTED ON THE PRINTED BOARD

Each type of component has one or more rework techniques best suited to its removal. Multi-lead devices such as plastic encapsulated chip carriers (PLCCs) are best handled by the more expensive hot gas jet, infra-red or heated electrode units, which remove the component in a single operation. However, such equipment is unnecessarily complex and costly for removing the simple chip resistor. The hot gas pencil and heated tweezers are much more suitable.

Very often the choice is made in response to factors other than the best option, for example tool availability, or designed-in proximity constraints.

(B) DEPENDENCE ON PRINTED BOARD LAMINATE TYPE

The type of printed board material used has two major effects on the choice of rework method:

1. For laminates with low copper peel strength such as PTFE, the tool and layout should enable sufficient visibility of the component joints to see that all are molten before lifting away.
2. For boards having high thermal mass such as metal-cored types or those with large-area ground planes, to avoid the use of a tool with a high heat input rate, the use of a hotplate to provide background heating is essential.

(C) DEPENDENCE ON ASSEMBLY STRUCTURE AND SOLDERING PROCESSES

Assemblies that have been wave-soldered carry devices that are glued to the board. In this circumstance, rework tools must be capable of supplying sufficient heat to melt the solder and soften the adhesive before applying lateral torque to twist the component and break the adhesive bond.

Assemblies without glued components, e.g. reflowed structures, do not require the latter facilities. It is sufficient to apply flux and melt the solder. The flux provides improved thermal coupling as well as reducing the oxide.

Where boards have surface mount components on both faces, control over the rework process must prevent damage to joints or loss of components from the reverse face directly opposite those being reflowed, as well as adjacent items. In some instances it may be advisable to design for the use of adhesive on one face even for reflowed assemblies. This will not prevent undesirable remelting of the joints and consequent intermetallic formation, but will at least prevent components from falling off.

Taking all these factors into consideration, the data in Table 16.1 indicate the recommended tool for use with a selection of typical component types.

Table 16.1 Matching rework tools to component types *(from original by M. Wickham)*

	SOIC	PLCC	QFP	LCCC	SOD	SOT	CHIPs	MELFs	Leaded Passive
Miniature conventional soldering irons	● ■ + ◄	(■ ◄) (● +)	(■ ◄) (● +)		● ■ ◄	● ■ ◄	● ■ ◄	● ■ ◄	● ■ ◄
					+	+	+		+
RF-powered soldering irons	● ■ + ◄	(■ ◄) (● +)	(■ ◄) (● +)		●	●	●	●	● +
Hot gas pencils	● ■ + ◄	● +	◄	◄	● +	● +	● +	● +	● ■ ◄ +
Heated tweezers				+	●	●	●	●	●
Modified soldering irons	●	●							●
Hot gas rework equipment	● (■ ◄) (+)	● (■ ◄) (● +)	● (■ ◄) (● +)	● (■ ◄) (● +)	(■ ◄) (● +) ●	(■ ◄) (● +) ●	(■ ◄) (● +) ●	(■ ◄) (● +) ●	(■ ◄) (● +) ●
Focused infra-red	● ■	● ■	● ■	● ■	● ■	● ■	● ■	● ■	● ■
Thermode equipment	●	●	●	●	●	●	●	●	●

● Component removal ■ Component replacement + Solder addition ▲ Solder removal ⬭ Not cost-effective

16.5 Preparation for rework and repair

(A) PRE-BAKING OF PRINTED BOARD ASSEMBLIES PRIOR TO COMPONENT REPLACEMENT

Prior to repair or rework, multilayer printed boards that have been returned from the field or have been in storage for a month or more should be baked. This is to reduce the risk of board delamination. The process should be carried out at the maximum storage or operating temperature of the assembly for an appropriate time. Typical combinations are 48 hours at 80 °C or 60 hours at 70 °C, depending on board size, layer count and component types on the board.

(B) ANTI-STATIC PRECAUTIONS

Suitable high-impedance earthing arrangements with circuit breakers should be made for all rework process equipment, benches and the workforce.

(C) REMOVAL OF ADJACENT COMPONENTS AND CONFORMAL COATINGS

Prior to any rework or repair operation, any components or parts that inhibit access to the terminations needing attention should be removed.

This can be a significant problem on boards that have seen service. Conventional components may have been assembled after testing, or surface mounted or programmable integrated circuits may have been socketed later and require extraction to allow access to surface mounted components beneath them. If board design is poor, this operation may necessitate desoldering of several items before the critical component can be worked on.

Boards returned for repair after service in the field may be conformally coated. This must be removed from the area surrounding the component before work commences, taking great care to avoid damage to components and substrate, e.g. by gentle use of airbrasion, chemicals or heat. Any debris or residue left from these processes must be thoroughly removed before rework starts.

Note: Some plastics can give off toxic fumes when heated to soldering iron temperatures, e.g. polyurethane varnish.

(D) CLEANING OF ASSEMBLIES PRIOR TO REWORK

For many tasks, the localized application of an isopropyl alcohol–water mixture applied by brush will be sufficient. Normally there is no need to replace the glue unless it was applied to give additional post-assembly mechanical strength—as opposed to holding components to the printed board during wave soldering.

If total immersion cleaning is required, ensure that all components and materials used on the board are suitable for such immersion and are compatible with the intended cleaning fluid.

(E) MASKING OF SENSITIVE COMPONENTS AND COATINGS
Where temperature-sensitive components or materials are close to the reworked component and hot gas heating systems are employed, it may be necessary to insert masking baffles or local screens—not an easy task.

(F) PREHEATING OF LARGE BOARDS
As well as shortening the rework process time, preheating is important in avoiding thermal shock and also in reducing the risk of delamination; see (A) above. It should be applied whenever practicable. For multilayer boards it must be considered absolutely essential.

(G) PREHEATING OF SENSITIVE COMPONENTS
Refer to 16.3(B) and 16.6(C2).

16.6 Rework activity classification

Apart from the necessary preparations, e.g. removal of conformal coating and anti-static precautions, there are seven basic elements in surface mounted component rework and repair activity:

- Component realignment ('tweaking')
- Component removal
- Addition of solder and flux
- Removal of excess solder from a joint or glue from the printed board
- Component replacement
- Cleaning (if required)
- Visual inspection of rework

Note: Post-rework cleaning and visual inspection are included in this list. The reason is that these activities contribute a major proportion of production costs in the rework loop and this should be exposed to management.

(A) COMPONENT REALIGNMENT
Even if there is no intention to lift away or remove the component, flux should still be applied to aid even thermal coupling and to enable a smooth joint free of spikes.

(B) COMPONENT REMOVAL
Typical defects raising the need for component removal include:

- Faulty component (electrical, mechanical)
- Expensive component (removal for future use)
- Component placed in wrong position or wrong orientation
- Design change
- Solder balls trapped beneath the component

Notes: Those surface mount components removed from boards should only under very exceptional circumstances be used again, and certainly not without 100 per cent visual inspection and 100 per cent retesting.

For example, multilayer ceramic chip capacitors exposed to undue thermal shock may develop internal microcracks during the removal and replacement heating operations. These may not be detectable at shipment and can fail three months later in the field. Also, chip ceramic capacitors, fixed chip resistors and trimmers have silver leached from their terminations during initial soldering operations. In any attempt to resolder them further leaching will occur, leading to depletion of silver at the ceramic interface and dewetting from the ceramic.

Where a high risk can be forecast of an integrated circuit needing removal, as in the case of memory expansion and programmable components, the use of sockets is recommended.

(C) ADDITION OF FLUX AND SOLDER
Defects likely to raise the need for addition of flux and solder include:

- Incorrect solder quantity—design or process fault
- Dry joint
- Solder theft, e.g. due to lifting of resist or poor printed board layout
- Required for new component

1 Applying flux
During all rework operations, the application of a good-quality liquid RMA or similarly mild flux is recommended. A mild or 'no clean' flux can be particularly important for repair when there may be no subsequent cleaning and it is essential to minimize corrosion risks from flux residues.

If flux is applied prior to component removal, often sufficient solder is left adherent to the pad to avoid the need for additional solder during replacement.

An even coating of flux can be applied with a cotton bud or soft brush. As well as removing oxides from the surfaces to be soldered, the flux layer also allows the solder to melt more quickly and evenly, thus reducing rework times and overheating risks. If preheating is used, the flux should not be applied until a few seconds before the rework action starts.

2 Applying solder

'Topping up' applies mainly to dry joints or those with insufficient solder supplied either by the assembly process or poor design or both.

Usually the choice lies between dispensing small amounts of solder paste and applying heat with a hot gas pencil, or applying flux to help the heat distribution before using plain solder wire with a small iron. When cored wire is used, some joints may not require pre-fluxing; others will, depending on their size and shape.

Reworking multilayer ceramic chip capacitors requires special care in the rework context. To minimize the risk of internal damage from thermal shock, the following precautions are essential:

- The chip should be preheated gently to 100 °C.
- If a soldering iron is used, its power rating should not exceed 30 watts, its tip should be no more than 3 mm diameter and its maximum tip temperature should be set to 280 °C.
- The maximum soldering time should be 5 seconds.
- Heat should be applied to the lead or termination area, never to the component body.

3 Starting almost from scratch

The 'almost' refers to the fact that after component removal the printed board footprint pads may have more solder on them than the original amount.

With dexterity and practice, solder paste can be dispensed from a 25 cl syringe on to individual pads prior to placing the new component(s); see 16.7(K)2 below. Tapered polypropylene nozzles are preferable to stainless steel tubes.

On multi-lead footprint arrays having a lead pitch of 1.27 mm (0.050 in) or less, it may still not be practicable to deal with individual pads, but the technique of laying the cream like a thin strip of toothpaste along the row of pads does work after a fashion. As heat is applied to the new component leads, surface tension pulls an amount of solder on to each pad, though a few shorts between pads may require attention.

Alternatively, with a miniature soldering iron in one hand and a pair of tweezers in the other, the iron can be applied first to two leads far apart to tack the component in position. The tweezers can then be discarded in favour of a length of solder wire and each joint worked in turn to provide the correct amount of solder to make a good joint. Great skill is required to make good-looking joints.

For fine pitch devices, e.g. at 0.635 mm (0.025 in) or less, it is normal to use a thermode machine; see 16.7(I) below. One of the best methods is to use a specially dimensioned kit of self-adhesive polyimide tapes with a solder strip which runs the length of the row of leads to be attached, for example Raychem 'SolderQuick'.

Careful alignment of the tape with the pads is essential prior to applying a thin coating of flux and then placing the component. Slots in the tape help in aligning the component leads, and the tape is carefully peeled away after soldering.

Heat can be applied using either a hot gas or a thermode machine. The former is sometimes preferred as, after placement, the component can be lifted above the pads so that the solder is visible while heated. When it is molten the component is then lowered on to the solder mounds and held until flow and then solidification occur. Focused infra-red (IR) machines can also be used for this process.

To reduce the risk of leaching, use of 2 per cent silver in rework solder is essential where ceramic chip components are involved.

(D) REMOVAL OF SOLDER
Comments on the performing of this task are given in each of the relevant parts of Section 16.7 below.

(E) COMPONENT REPLACEMENT
Defects raising the need for component replacement are:

- Component missed during placement
- Component dislodged during soldering or cleaning
- Component removed during rework
- Component not available at time of initial assembly
- Design change

Comments on the performing of this task are also given in each of the relevant parts of Section 16.7 below. Where adhesive has been used to attach components, a suitable solvent or heated blade will be needed to remove sufficient glue residue to allow each replacement component to be seated on appropriate solder pads.

16.7 Rework machines, tools and methods
This section covers the principal techniques only.

(A) MINIATURE CONVENTIONAL SOLDERING IRONS (MANUAL)
These irons have small tips and lower thermal mass to enable safer reworking of the finer geometries found on surface mount boards and components.

The use of temperature-controlled versions is essential, e.g. at 260 °C ± 20 °C, with regular checks on their bit temperatures as they are prone to major variations while in use.

1 Component removal

Leadless ceramic chip capacitors and resistors are removed by applying the tip of the hot iron to the centre region of the chip (for removal and disposal only) and, when both joints are reflowed, using a pair of fine metal tweezers to twist and lift it away.

The procedure must be regarded as destructive for the components. On no account should they be re-used. Although this method does work, it is not the recommended one for passive chip components.

Packaged devices with tape or wire leads are removed with difficulty by reflowing each joint in turn and bending each successive lead up and away from the printed board until all are clear. A vacuum pencil or large aperture tweezer is used to lift out the component body. This approach is also destructive, very time-consuming and likely to be impracticable for PLCC packages. In some cases it may be necessary to precede the main operation by removing excess solder from individual joints with a miniature vacuum soldering iron or with copper braid.

2 Solder addition

For addition of solder, the use of cored or solid wire is recommended.

3 Solder removal

As stated above, use clean, good-quality fluxed copper braid.

4 Component replacement

This tool is not recommended for replacement (or, indeed, initial soldering) of chip capacitor components except in prototype situations where product reliability is not important. Severe invisible internal damage can be caused to components by inexperienced operators, of the type that may become catastrophic after several months of operation in the field (see 16.6(c2) and Section 17.2).

(B) RF-POWERED SOLDERING IRONS (MANUAL)

These irons incorporate small heaters designed into the tip and are powered from a constant current source. The tip is heated by skin effect and, being self-regulating, draws power only when it is required. However, like conventional small irons, they are capable of causing rapid heating and should not be used for replacing or touching-up ceramic chip capacitors.

Soldering can be carried out satisfactorily at lower tip temperatures than with conventional irons, making rework less operator-sensitive and reducing the risk of thermal damage to board and component.

Radio frequency (RF) irons are much quicker to heat up and cool down, allowing tips to be changed more easily. Because of the self-regulating feature, one tip can be used for a wide range of components and printed board thermal masses. However, for best results with surface mount devices, a range of specially designed tips is available for up to 68-pin packages. Some users report that larger tips are less effective than the smaller ones.

The main disadvantage of RF-powered soldering irons is the comparatively high cost of a work-station, e.g. £500. Also, the health and safety aspects should be carefully investigated.

(C) HOT GAS PENCILS (MANUAL)

Hot gas pencils are small, hand-held instruments which dispense a controlled stream of heated air. This jet is fine enough to be directed at individual joints and will supply enough heat to reflow them, one by one. They can be used for all types of rework, and portable versions powered by Butane gas to heat the air are also available.

1 Component removal

For leaded components, each individual lead has to be reflowed and bent up clear of its pad using tweezers to prevent the joint reforming. When all leads have been treated this way, the device can be lifted away with tweezers or a vacuum pencil.

For multi-lead component types, as with the miniature soldering iron, the method is very slow, but is possible where other more suitable techniques are not available. However, it may not always be able to remove large leadless components such as 2220 size chip ceramic capacitors, and certainly pencils cannot manage LCCCs.

When used on smaller devices such as ceramic chips and SOT 23 device packages, it is possible to melt all terminals at once by playing the jet to and fro. The gas spreads out sideways as it hits the board and care must be taken to avoid overheating adjacent components. If there is sufficient heat to remelt the joints to small chip components, they may be blown sideways off the board by the jet.

2 Solder addition
Use of solder paste rather than wire is recommended.

3 Solder removal
Use in conjunction with good-quality fluxed copper braid on individual joints.

4 Component replacement
Replacement can be achieved by first dispensing solder paste on to the footprint pads using a pressurized syringe, then hand-placing each component and either reflowing each joint in turn or oscillating the jet to reflow all joints on small components simultaneously.

For multi-lead devices the method is very slow, but, again, is possible where other more suitable techniques may not be available.

(D) HEATED TWEEZERS (MANUAL)
Similar in principle to solder guns, these tweezers have specially shaped resistive metal heads which carry current to generate heat at their tips, but this current does not pass through the reworked device.

By gripping the ends of a two-terminal component with the tweezer tips, heat is transferred to the joints to reflow them simultaneously so that the body can be lifted clear before the solder cools and the joints reform. Its main advantage is that it offers a one-handed operation.

Heating can be via pulsed or continuous-current flow. The tool is particularly suitable for reworking glued components which need to be rotated through 90 ° to shear the adhesive after the solder joints become molten. A drawback of this tool is that it can be difficult to see exactly when the solder becomes molten, and there is a risk that an inexperienced user may act too soon and peel the copper pads away from the board during the twisting action.

The printed board layout should be designed to allow room for this action.

1 Component removal
This is suitable for small two-terminal components and SOT 23 types only. After placing flux on the joints, the action is as described above.

2 Solder addition and removal
Touching-up, solder addition and solder removal are not practicable with this type of tool and should be carried out using an alternative method.

REWORK AND REPAIR 195

3 Component replacement
This tool is not generally recommended for component replacement of temperature-sensitive chip capacitors because of thermal shock.

(E) MODIFIED SOLDERING IRONS
These are conventional soldering irons fitted with special interchangeable tips to suit a wide range of standard surface mount device packages. Currently they are available to fit all standard ceramic chip components (resistors and capacitors), MELFs, SOD/SOT packages and SOICs.

Sometimes users attach them to small drill stands to ensure better control of downward pressure and to maintain coplanarity of the tool face with the printed board. A tip temperature of 260 °C is recommended.

Some manufacturers offer tips for the smaller PLCC sizes—up to 44 pins. The tips can be used to apply torque to glued components when the solder is molten, but this latter condition may be difficult to see, depending on the depth of the tip cavity and the proximity of adjacent components. Particular care is needed when working on printed board materials having a low copper peel strength.

1 Component removal
The better-quality tool heads are equipped with a vacuum chuck to aid lifting of the package when reflow temperature is reached.

2 Solder addition and removal
Solder addition and removal are not practicable with this type of tool and should be carried out beforehand by an alternative method.

3 Component replacement
This tool is not usually recommended for component replacement, but may be used in emergency if nothing else is available. A considerable amount of touch-up may be required.

(F) HOT GAS REWORK MACHINES
Note. Hot-air 'DIY' paint strippers, heat shrinking guns and hand-held hair dryers should be avoided.

Hot gas rework machines are benchtop equipment capable of dispensing a stream of heated gas (generally air, but sometimes nitrogen) through nozzles or via a baffle block on to a printed board and component. They are intended mainly for use with multi-lead packages (see Fig. 16.1). The nozzles generally form a shaped array of small tubes set in a block and bent near their tips to direct the flow towards the component joints.

(a) Multi-jet type

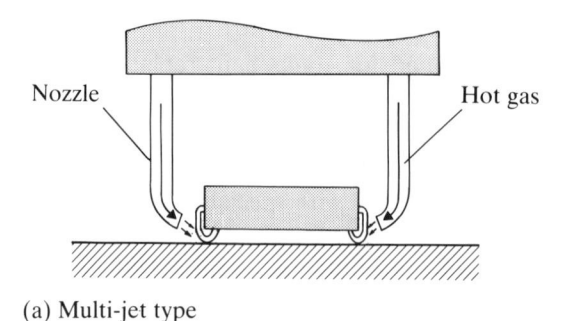

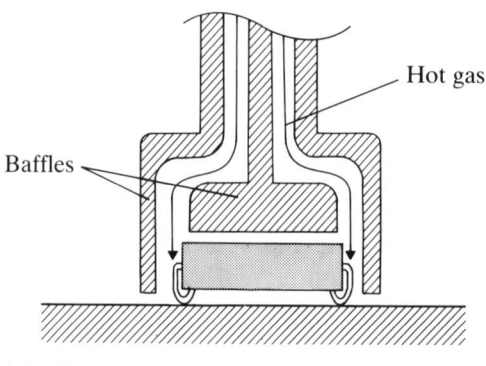

(b) Baffle type

Fig. 16.1 Basic hot gas rework machine (*from original by M. Wickham*)

A different set of nozzles is required for each integrated circuit type, but on some machines the array for a large device will also rework smaller ones—provided the printed board layout design has allowed enough room around the package to permit access.

Nozzles are interchangeable, but not rapidly. It can speed things up if unheated air is allowed to pass through the nozzles to cool the block. More sophisticated machines have a single array of fine nozzles which can be made to 'open' and 'shut' by remote control, thus giving a programmable jet-stream contour to suit different package types and sizes.

Most machines have an integral hotplate or a secondary jet mounted underneath to provide preheating. In volume production it is advisable to purchase an auxiliary hotplate or heated tunnel with a tapered temperature profile so that preheating begins earlier and thermal shock is avoided as the boards are moved towards the rework station.

The gas temperature and flow rate are regulated and these determine the time to reflow. A thermocouple is located in the jetstream and this controls the electrical power in the heating element. Too high a temperature–flow-rate combination can cause reflow of joints on nearby components. Many machines also include a timer to shut off gas flow after a definable period to prevent overheating of the component and the printed board.

The process can be viewed through a binocular microscope or alternatively via a video camera and monitor screen. Most systems have an integral vacuum chuck which can be activated to lift the component off the board, and some units are fitted with sensors or springs attached to the chuck which will enable it to lift automatically when the solder is molten at all the joints.

Several versions are available with an auxiliary small single-jet pencil for use on individual joints.

This type of equipment is one of the best options for removing and replacing high-pincount devices.

1 Component removal

Liquid flux should be applied to all joints and then all should be reflowed almost simultaneously using the correct nozzle before lifting the component body away with the vacuum chuck provided.

2 Solder addition and removal

Unless the equipment is fitted with a single-jet pencil, it cannot itself be used to apply additional solder or to remove excess without reflowing all joints. Pre-placement of solder paste, fluxed preforms or pre-coated polyimide adhesive strips are usually required for component replacement. If a pencil is available on the machine, for solder addition its use in conjunction with solder paste is preferred to the application of wire with a soldering iron; see 16.6(c).

3 Component replacement

After fitting the correct nozzle, an appropriate amount of solder and flux is placed on to each footprint pad using one of the methods given in 16.6(c) and reflowed as appropriate. The component is then loaded into the nozzle and positioned over the refluxed footprint pads, using the viewer to assist alignment. The nozzle is lowered gently into place and heat is applied. On some machines it may be necessary to raise the component a small distance above the board before applying heat, depending on the level of preheat imparted to the board.

In the latter case, when the solder is molten the component is lowered and flow will occur after a few seconds. Heat is switched off and the chuck is held in position until solidification is certain. Joints are reflowed almost simultaneously.

(G) FOCUSED INFRA-RED LIGHT BEAMS

This type of equipment consists of a focused beam of short-wavelength infra-red light, collimated or shuttered to restrict its area to suit the component joint(s) being reflowed (see Fig. 16.2).

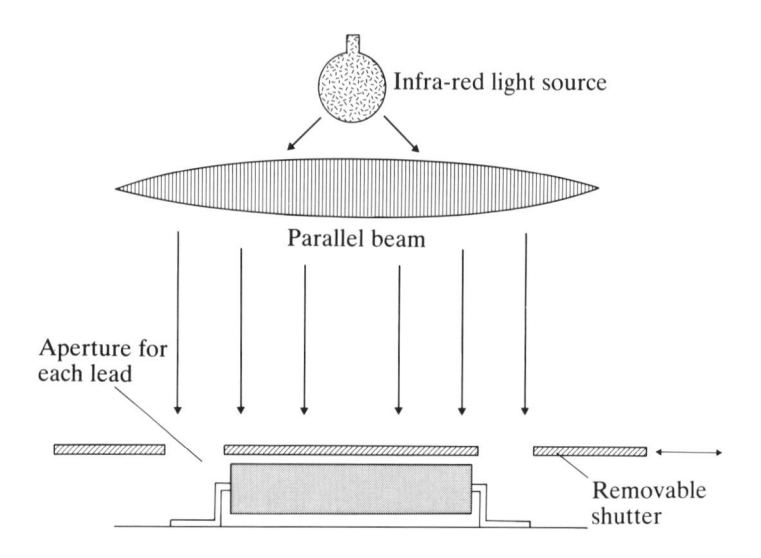

Fig. 16.2   Basic infra-red rework head

Owing to the method of heat transfer, the process is comparatively slow and black plastic mouldings can get very hot unless suitably apertured shields are employed or great care is used in setting up and programming. Obviously the latter point is unimportant if the component is to be thrown away after removal.

To reduce the process time, either the machine should be fitted with a programmable preheat station, or the board should be preheated away from the machine and transferred quickly thereto.

Work is placed on a platform capable of x, y movement and the heat cycle can be programmed to provide different time–temperature profiles for component terminations of differing shapes and thermal masses. More sophisticated versions have split-beam optical systems to facilitate alignment of fine pitch integrated circuit package leads with the printed board footprint pads and quickly changed shutters bespoke to individual package types and sizes. These direct the infra-red beams only at the leads, not at the embedding plastic.

Additionally, there can be an integral vacuum pick-up head which reduces the likelihood of lifting printed board tracks by repeatedly trying to lift the component away using just enough force to break the surface tension of the molten solder.

1 Component removal

Prior to removal, liquid flux should be applied to all joints. Although the shutter and thermal programme should be similar each time a specific package type is reworked, with different printed board thicknesses and layers, several trial runs for each new circuit may be necessary to establish the right time–temperature programme.

2 Solder addition and removal

Unless the equipment is fitted with a miniature soldering iron or a single-jet hot gas pencil, it cannot itself be used to apply additional solder or removal of excess without reflowing all joints. Pre-placement of solder paste, fluxed preforms or solder-coated polyimide adhesive strips are usually required for component replacement. If a pencil is available on the machine, for solder addition its use in conjunction with solder paste is preferred to the application of wire with a soldering iron (see 16.6(C)).

3 Component replacement

Method A. An appropriate amount of solder (in paste or other suitable form) is placed on to each footprint pad using one of the methods given in 16.6(C)2 and reflowed as appropriate. The solder mounds are refluxed and a new component is then placed in the machine and lowered on to the footprint pads, using the viewer to assist alignment. The correct shutter is inserted and the heat programme initiated. Flow will occur after a few seconds and the component will settle on to the pads under the weight of its chuck. The heat is then switched off automatically and the chuck held in position until solidification is certain.

Method B. With extreme care in alignment, a suitable self-adhesive polyimide strip pre-coated with solder is placed on the printed board where the component is intended to be mounted. Flux is applied to the strip and then both the board and the replacement component are loaded to the machine.

The correct shutter is inserted and the correct preheat and reflow programmes are initiated. Flow will occur after a few seconds and the component will settle on to the pads under the weight of its chuck. The heat is then switched off automatically and the chuck held in position until solidification is certain.

(H) THERMODE OR HEATED ELECTRODE (HOT BAR) EQUIPMENT

This method is used for initial component placement and soldering as well as for rework.

Developed from small welding machines, thermode equipment can be used equally for soldering and desoldering. They apply pulsed resistance heating methods to shaped electrodes which are designed to make simultaneous contact with all the terminations of multi-lead devices. The electrode materials are not wettable by the flux–solder combination used on printed boards, and, as with heated tweezers, virtually no voltage appears across any of the device terminations during the operating cycle (see Fig. 16.3).

The electrode system for rework is usually mounted on a form of dieset in

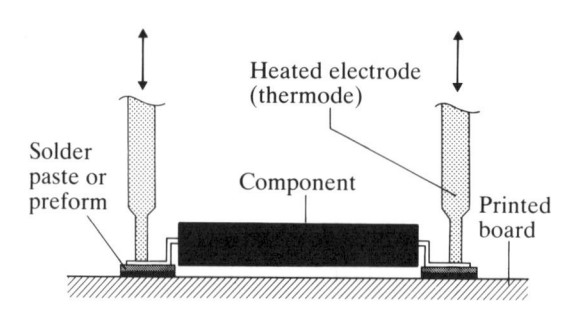

Fig. 16.3 Basic thermode rework tool head

order to give very accurate vertical motion and to maintain the electrode face coplanar with the dieset base and the printed board surface.

It is important to check coplanarity on a regular basis, e.g. hourly, and to ensure that the jigging arrangement for each individual printed board footprint location allows the dieset to control the coplanarity of the printed board at that location with the base and the heated electrode. The more sophisticated systems have a floating head which can follow the surface of the printed board and thus ensure that device leads are held flat on to its surface.

Thermode technique is particularly suitable for quadpacks and flat-packs—which, incidentally, were designed for this method rather than for the reflow processes now applied to them (hence the problems). However, with an accurately shaped electrode to suit each particular supplier's product, PLCCs can also be removed and replaced with a thermode.

The removal and replacement process cycles can be time-consuming because of the need to heat comparatively large thermal masses, and in the latter operation the electrode has to remain in place holding the leads hard on to the printed board until the solder has solidified. For both removal and replacement, board preheating is advisable up to 100 °C.

For lead spacings down to 0.8 mm, use of solder paste for replacement is practicable, but there may be some shorts to rectify, e.g. with a miniature soldering iron. To avoid contact with surrounding components, the paste will have to be dispensed from a syringe. Below that level of lead spacing it is difficult to position paste evenly on each pad in small enough quantities, so one possibility is to use a low solder content paste instead of the usual type. Alternatively, there may well be sufficient solder remaining on the pad after component removal, in which case only the addition of flux is needed.

As with hot gas equipment, thermode units usually have a vacuum chuck built into the electrode system for lifting the component away after reflow.

1 Component removal

Apply a small quantity of RMA-type flux evenly with a brush or nozzle just before the board is loaded beneath the rework head, i.e. after preheat. Lower the electrode to the leads and apply power to the heating electrode. When the solder is molten, activate the vacuum chuck and raise the electrode with the component held within.

2 Solder addition

The method should be confined to situations in which all the joints require the same amount of solder added for mounting or remounting the component.

As stated above, dispensing solder paste beforehand is the preferred method in most cases. In special circumstances, the use of fluxed solder preforms may be a suitable alternative.

3 Solder removal

Solder removal is not part of the function of this type of equipment and should be carried out beforehand by an alternative method.

4 Component replacement

After adding solder paste to the footprint pads, the replacement device is either positioned on the printed board by hand or located within the electrode and held up by vacuum. The electrode is held in position during heating and cooling cycles and upon their completion is lifted clear.

(I) LASER DESOLDERING

Use of the laser as a rework tool cannot yet be recommended.

(J) CONVENTIONAL (E.G. 50 WATT) SOLDERING IRONS

Very great skill is needed to avoid damage via thermal shock to surface mount components. Such large irons are definitely not recommended except for the larger sizes of MELFs or through-hole components adapted for surface mounting. In any case, they should be avoided on printed boards where individual components are spaced with gaps less than 6 mm.

(K) OTHER REWORK TOOLS

1 Hotplates

The use of hotplates to provide background heating before and during rework is strongly recommended wherever it is practicable. In addition to reducing the energy needed from the rework heating tool, background heating also dramatically reduces the risk of thermal shock to the component to be reworked, to components surrounding it and to the printed board itself. Preheat is particularly important when working on multilayer boards where moisture ingression between layers can cause delamination if insufficient precautions are taken.

2 Pneumatic dispensers

Pneumatic dispensers provide a consistent, controllable source of solder paste which is ideal for the rework environment. Most units are fitted with an air pulse pressure regulator, a programmable pulse length timer and a gauge to indicate the pressure from a normal oil-free factory compressed air supply. The pulse is triggered by a foot switch.

When connected to a syringe of solder paste, the pressure can be adjusted and the length of pulse programmed to control the volume of paste dispensed by one depression of the foot switch. On simpler equipment the pulse time is not controlled automatically.

These units dispense solder paste with significantly better control over quantity than hand-held syringes.

3 Desoldering tools as used for conventional through-hole assemblies

Desoldering tools are of limited value for removing excess solder from surface mount joints or desoldering surface mounted components with low pincounts because the sizes of the bit tips tend to make them too cumbersome. However, they are extremely useful for removing excess solder from footprint pads after defective components have been lifted off the printed board. Some types have a vacuum pick-up nozzle to suck away the excess solder.

4 Tweezers and vacuum pencils

The choice between tweezers and vacuum pencils should be left to individual operators wherever possible.

Any tweezers used should be of the soft-nosed, non-tinnable variety, e.g. high-temperature plastic or bone, to avoid damage to components. Preferably they should also be conductive to avoid having to differentiate between static-sensitive components and others.

Vacuum pencils may be preferred for delicate components, but not all devices have the flat upper surface necessary for pick-up. Trimmer resistors and capacitors are examples. Some vacuum pencils have the additional feature of a rotatable head which can be useful when aligning components during hand placement.

Both types should be kept scrupulously clean, It is advisable to issue formal instructions which call for cleaning on a routine basis and specify the method, e.g. at least hourly, or more often if contamination of any sort is observed.

16.8 Rework recording procedures

Important reasons for strict control of rework procedures are the collection of data for cost improvement and the control of product status.

Where inspectors are doing the rework, they should be given a Rework Chit with each board. This would already carry the board number (if logged) and the batch number. Inspectors are asked to insert the date of inspection and the number of defects in each category. Their work should be checked by a different inspector.

If the rework requirement on a particular board has already been identified (e.g. arrowed) and logged on a Rework Chit by an inspector and passed to an operator to carry it out, it is important that the paperwork enables the same circuit to return with its chit to the same inspector, preferably with the arrows still in place; otherwise the inspector may have to recheck the whole board instead of just the reworked joints.

Some will argue that a complete recheck is no bad thing, because it is widely accepted that for surface mount assemblies visual inspection is not an efficient method of finding all faults and operator fatigue can seriously affect results.

To ensure that its status is known, a label or coloured paint mark which survives the cleaning processes should be applied by inspection to every board that fails visual inspection. For each unsuccessful rework cycle, the board then receives a separate penalty mark of the same colour, until it passes and receives a mark of a different colour—thus indicating its history in a low-cost manner.

The same effect is obtained by attaching a label to each board and marking this, but the label must also be able to survive the necessary processes, and it invites a slightly higher chance of errors in status identification because it can be removed.

Different colours are used to indicate different stages after rework, e.g. failed visual inspection, passed visual inspection, failed in-circuit test, passed in-circuit test, failed functional test, passed functional test, failed final visual inspection, passed final visual inspection, etc. Boards that required three rework cycles would carry three marks of the same colour, e.g. on one edge designated for the purpose.

This is one of the most effective ways of building both a feedback system and a realistic actual cost structure, especially for large complex printed board surface mount assemblies.

The difference from previous systems is that failure to pass is now treated as worthy of marking, and this provides a simple means of generating more pressure for corrective action. When a manager looks at assembled boards at the very end of the line, the situation is made clear at a glance. Where there is no room on the board surface for such multiple markings, the edges can be used.

If it is considered inappropriate to dispatch marked boards, a method of removing them without harming the circuit must be found, but the loss of data should be agreed by senior management beforehand. Alternatively, the use of tie-on labels may be preferred.

16.9 Field repair philosophy

Consideration should have been given during the printed board layout design stage to policy on maintenance of products that contain surface mount assemblies and are not considered to be 'throw-away' items.

Great harm can be done by untrained and ill-equipped repair workers in the field, not only to the circuitry, but also to the company's reputation and brand image.

Wherever practicable, surface mount circuits ought to be returned either to the original assembly line, where the necessary experience and test gear to carry out repair work specific to the circuit should be available, or to specialized skill centres.

17

Surface mount field reliability issues

17.1 Introduction

It is vital that OEM staff responsible for placing orders for outside assembly and for assessing subcontractors should understand the main reliability problems that are seen when using surface mount technology in the field.

The reason for attaching such importance to the subject is that many of the problems seen have arisen through subcontractors' lack of knowledge, leading to unwise purchase and use of soldering equipment. Very small subcontract firms have been more likely to experience these difficulties. This stricture applies equally to rework which may be carried out after assembly or, indeed, at any time during the life of the product.

However, the relatively few major disasters in applying surface mounting usually have had the additional elements of poor board layout or assembly structure design, or ill-advised component choice—none of which can be laid at the door of the very small subcontractor.

Readers are advised to ensure that the following points are drawn to the attention of managers in circuit design, printed board layout, purchasing, quality and production engineering teams. The recommended protective measures suggested below all cost money, and it is up to the relevant department heads to advise top management on whether to afford the 'insurance premiums', or to take the risk of not doing so.

Apart from fatigue failures, which can occur in any set of component solder joints after prolonged thermal cycling (Section 17.4), the two components giving trouble most often have been multilayer ceramic chip capacitors (Section 17.2) and large multi-pin integrated circuits in plastic packages (Section 17.3).

17.2 Multilayer ceramic chip capacitor failures

(A) BACKGROUND

There continue to be instances reported in which multilayer ceramic capacitor batches from some suppliers have developed short-circuits after mounting. Over the five years since 1987, I have been aware of cases involving fire, litigation, arbitration, and replacement with wire-ended types, all arising from this problem. The internal structure of a typical multilayer chip capacitor is shown in Fig. 17.1.

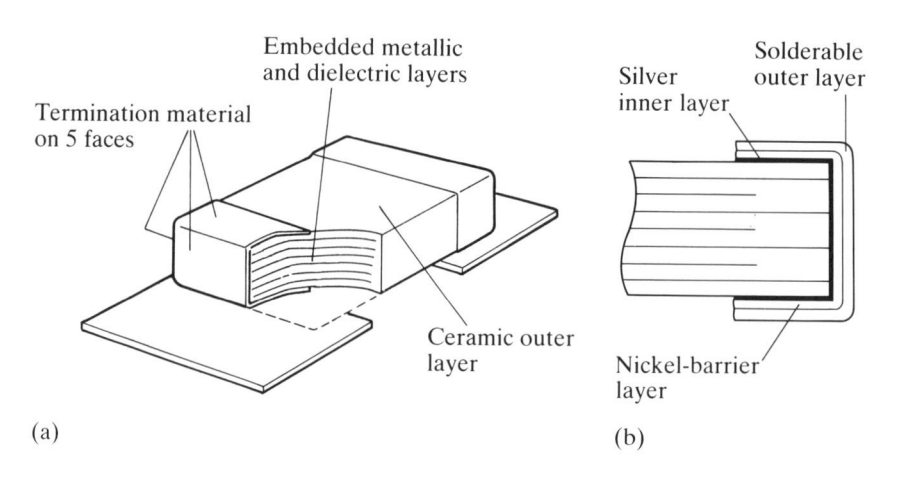

Fig. 17.1 Typical multilayer ceramic chip capacitor

Often the defects are not detected during final test and failures occur after several months in the field. It is known that the main cause is internal breakdown arising from micro cracks in the dielectric which immediately or, more often, eventually allow normally separated metallic electrode layers to make contact at a later date.

Some sources say that this internal breakdown phenomenon is not entirely unique to chip versions of multilayer capacitors and that it also occurs in equivalent leaded types. However, there can be no doubt that the higher process temperatures, greater thermal shocks and more severe mechanical stress levels seen by surface mounted versions must raise the risk levels compared with wave-soldered wire-ended components.

Market pressures demanding increased miniaturization via more microfarad-volts per cubic centimetre have encouraged some manufacturers progressively to use more and thinner dielectric layers. Alongside the introduction of the nickel barrier to reduce solder bath contamination rates and leaching effects, this has lowered the threshold of sensitivity to thermal shock for some size combinations.

In fairness to suppliers, it must be pointed out that, although failures in mounted chip capacitors can arise from component design, material or multilayering defects, they have also been attributable to incorrectly printed board layout design and/or unsuitable assembly processes. The latter occur more frequently when the soldering equipment is incapable or has little margin in delivering the required thermal profile.

Conversely, to protect themselves from embarrassing legal claims, some manufacturers have specified thermal profiles that are barely practicable for the assembler and which can introduce other potential hazards to circuit integrity. For example, the requirement to preheat each chip to within 100 °C of the solder bath temperature at the instant of hitting the wave may cause harm to some types of conventional through-hole components present such as electrolytic capacitors, inductors, and even the printed board itself. (See also (D) (iv) below.)

In several instances it has proven extremely difficult to pin down the cause of chip failure because of a lack of recorded data or lack of awareness of the best way to proceed when such defects are reported. This chapter is intended to provide guidance in minimizing the risk of encountering the problems and in investigating them when they occur.

(B) CIRCUIT DESIGN

(i) During the electronic circuit design and component specification stages, engineers should consult with suppliers on the intended soldering process type and the consequent assembly time-temperature profiles planned for the components. This is also good practice from the product liability standpoint.

(ii) The temptation to specify the smallest available chip size for a given capacitance value should be resisted. More instances of failure have been seen in 100 nF 1206 chips than in any other, although this may be due in part to the fact that it is a very widely used value for decoupling purposes. In this context, exchanging reliability for smaller size is bad design practice.

(iii) Circuit designers may not generally be aware that, for a given multilayer chip ceramic capacitor value, printed board material and assembly process, there is a need to get the best compromise between chip dielectric thickness, overall chip thickness, lateral dimensions and metallization type. In all cases, uniformity and homogeneity of metal and dielectric layers remain key factors. Mistakenly, some designers tend to regard these things as the responsibility of production, but this attitude is considered to be non-professional.

(iv) Thicker capacitors are less likely to suffer internal fracture from mechanical stresses, but if their thickness has embodied a large number of layers, they are likely to be more sensitive to thermal shock.

Conversely, very thin chips are more prone to mechanical damage, typically from the centring tweezer jaws on some pick-and-place machines and from board flexure. Damage can arise from flexure during electrical test. See also (viii) below.

(v) Nickel barrier versions are known to be slightly more sensitive to thermal shock because of the increased metal content on the termination surfaces. When heat is applied rapidly, the metal heats up faster than the dielectric and sets up the aforementioned strains which can lead to microcracks within the capacitor structure.

Some manufacturers are beginning to offer alternative metallization combinations for use where wave-soldering heat cycles are short and well-controlled.

(vi) Occurrences of thermal shock failures are more often reported after wave- and vapour-phase-soldering than after infra-red soldering, but the latter method is not immune.

(vii) Responsible manufacturers are now marking in their catalogues those capacitor size–value combinations that should not be wave- or vapour-phase-soldered. These tend to be the top capacitance values in any given plan size.

(viii) Once assembled on to epoxy–fibreglass printed board material, both larger-area and thinner chip capacitors are more prone to damage from board flexure and from stress at low temperature arising from thermal expansion/contraction mismatch between the capacitor body and the printed board. Designers should avoid their use on FR4 printed boards that are likely to experience mechanical flexure or are intended for storage or repetitive operation over wide temperature ranges.

In this context, regardless of chip size, surface mount printed boards carrying ceramic components should not be distorted from the shape they assume after soldering, e.g. by subsequent handling in production, testing, installation or use. Opportunities for internal damage come from a wide variety of mechanical, thermal and thermo-mechanical sources, typically:

- At break-out from a panel containing step-and-repeat circuits
- On test fixtures in which a vacuum system or an array of probes is allowed to bend or distort the board
- On thin, rack-mounted boards whose bow is straightened every time the board is pushed into the rack or forced into a stiff plug/socket on the backplane
- Accidental dropping of the board
- Installations which exert cyclic stresses on the board, e.g. when a large bowed board is mounted rigidly with screws on to an aluminium frame and the equipment is switched on and off daily

FIELD RELIABILITY ISSUES 209

(ix) Derating by specifying capacitors with a rated voltage well above the maximum operating voltage seen in use is good practice and may slightly reduce the risks. However, this approach can no longer be accepted as giving adequate protection on its own.

Where power rail decoupling is involved, ensure that fail-safe circuit design is incorporated to prevent hazardous circumstances arising, e.g. risk of heat damage and fire from capacitor short-circuits. Greater safety margins can be obtained by specifying approved suppliers known to have well-controlled manufacturing processes and who do not produce components that have the thinnest dielectric layers.

(x) Check that printed board layouts do not encourage the formation of excess solder mass at chip capacitor terminations. Excess can arise when small or large pads are given too much solder paste in the printing process prior to reflow, or from wave-soldering with a bath temperature that is on the low side. Poorly controlled rework can also result in too much solder being present.

(xi) Because of the opportunity for better control of solder quantity and joint contour in chip capacitor attachment, many users prefer to design for reflow-soldering when their circuits are likely to see wide temperature ranges in storage or operation. In extreme cases, users specify chip capacitors fitted with metal feet (if available).

(xii) Ensure that process information given in supplier's data sheets for all specified components is passed on to production.

(C) PRODUCT APPROVAL TESTING AND GOODS INWARDS INSPECTION

(i) Qualification testing of ceramic multilayer capacitors is usually carried out after mounting by exercising them at low voltage via a high impedance source; 100 hours at 85 °C in 85 per cent relative humidity using 1.5 volts through 150 000 ohms is a typical schedule.

(ii) Several large companies experienced in volume SM assembly claim that there are only a few manufacturers capable of providing reliable chip capacitors. It has been established that such users have different opinions of the same suppliers. This and other information suggests that the reported problems are more likely to be batch-related than some suppliers will admit. The screening and batch test protocols adopted by different manufacturers should be studied when selecting suppliers.

(iii) Goods inwards inspection should include visual examination of sample capacitor bodies for delamination (local discoloration is a clue), leaching of conductor material, and surface cracks and chip-outs, neither of which should expose buried conductor material. The above checks do not guarantee avoidance of failure from thermal shock, but they are still worthwhile for cost-saving reasons.

(D) ON THE PRODUCTION ASSEMBLY LINE

(i) Check that the inward acceleration/velocity and gripping force of the pick-and-place machine jaws are suitable for the purpose and that jaw faces are kept clean and free from excessive wear. Some machines enable changes in the programme to allow for more sensitive components. Small pieces of ceramic detritus lodged in the jaw faces can exert very high local pressure on the chips and start micro cracks.

Machines employing automatic alignment via optical correction systems do not necessarily use jaw centring methods and the risk of damage to thin chip capacitors is reduced.

(ii) Bowed boards and incorrect machine settings can cause excessive downward placement forces on some pick-and-place machines. These may initiate cracks in large plan area and/or thinner ceramic capacitors.

(iii) Check the supplier's data sheets for guidance on thermal exposures.

(iv) Make sure that thermal shock during assembly soldering is minimized. Reasonably safe advice is to keep all heating and cooling rates below 2 °C/sec.

When wave-soldering, the preheat should ensure that the difference in temperature between the chips and the solder bath temperature at the instant of immersion is less than 100 °C, though some suppliers specify 70 °C. This means that, if the solder bath is maintained at 240 °C, the preheat temperature must be higher than 140 °C, which is above the glass transition temperature (Tg) of FR4 printed board material. Large printed boards may need support to prevent collapse, 'submarining' and permanent bow.

Look out for wave-soldering machines that have a large gap (>2 in) between the end of the last preheater and the start of the wave—it can mean that there is actually an undesirable drop in component temperature before hitting the wave-front, thus increasing the thermal shock level. At the point of immersion, it is impossible to meet the 2 °C/sec requirement.

(v) Check the thermal profiles seen by chip capacitors across each board design. Use conventional reeled thermocouple rigs or a 'Mole'. Attaching thermocouple tips direct to the printed board is not always good practice as, owing to the thermal mass of the board, they may indicate less severe conditions than are seen, for example, by small chip components. For the same reason, their attachment to such components will also show a thermal profile different from reality because of the comparatively large thermal mass of the thermocouples.

A few degrees extra should be allowed to cover these discrepancies. Suitable margins can be determined only by experiment, e.g. by comparison with nearby unattached thermocouple tips held just away from the board surface.

Sensors built into the soldering equipment are not suitable for defining profiles seen by components, but may be useful as comparators to monitor the consistency of the process. Bear in mind also that reflow profiles are highly dependent on the spacing between successive circuit panels laid on the belt.

Ignorance of what is happening to components during soldering may cost the company a lot of money further down the road.

(E) REWORK
(i) The safe advice is: 'Never re-use any chip capacitor after it has been mounted and removed, regardless of the method of desoldering applied.' The component manufacturers are unlikely to accept such items as defective under warranty, and by analysing the termination material after the necessary two extra remelts, they can usually detect when resoldering has occurred.

(ii) Ensure that chip capacitors are never touched-up or replaced using a soldering iron. Essentially an in-contact tool, the soldering iron is an established source of thermal shock and should be applied only for removal prior to discarding the chip. The preferred method of reworking chip capacitors (and all small discrete components) is to use a hot gas pencil.

(F) WHAT TO DO WHEN THE PROBLEM ARISES
(i) Do not remove the chip from the board immediately. The supplier can often better diagnose the cause by examining the chip while it is still mounted. Removing it may destroy valuable evidence.

(ii) If it is essential to remove the chip, make certain that the top side is marked with a material that will remain legible during desoldering. The manufacturer may be able to determine whether the cause was thermal shock or has arisen from board design or accidental mechanical damage.

(iii) Notify the chip supplier, but also ask an independent consultant or laboratory to examine some of the defectives. When the problem is of a very specialized nature, it may well be that the only source of a third-party opinion is a competitor-supplier. The original supplier will not be happy about this, but sometimes the offer of an unedited copy of the report is sufficient to help things along—especially if scientific knowledge is advanced for all parties.

(iv) Check the relevant soldering machine thermal profiles. The supplier may want to do this even if you do not or cannot do so yourself.

(v) Impound any part-filled reels or vials from the supplier that were (or may have been) used for the product concerned. It is good traceability practice to be able to correlate product batch numbers with component purchase order numbers, but this discipline does cost more to run than a freewheeling system and may it fail if part-used reels are allowed to remain on the shop-floor after a production run.

(vi) If appropriate, activate customer notification and product recall procedures.

(vii) If no redesign is necessary, order fresh supplies and plan the rework exercise. Make sure that all replacement chips are soldered using a hot gas pencil rather than a soldering iron. Failing to act on this may result in more trouble from the field in a few months' time.

17.3 Failures in large semiconductor integrated circuit plastic packages

The results of the failures described below are known colloquially as the 'popcorn effect'. This arises when devices have received enough exposure to humid atmospheres to allow water absorption and consequent moist air traps within the plastic encapsulation.

Figure 17.2 illustrates the nature of the failure. Upon sudden heating, the moist air turns into steam and exerts sufficient internal pressure to crack open the plastic. The cracks often occur beneath the device and when this happens they are not obvious during visual inspection. Failure can occur months later in the field as a result of ingress of contaminating material.

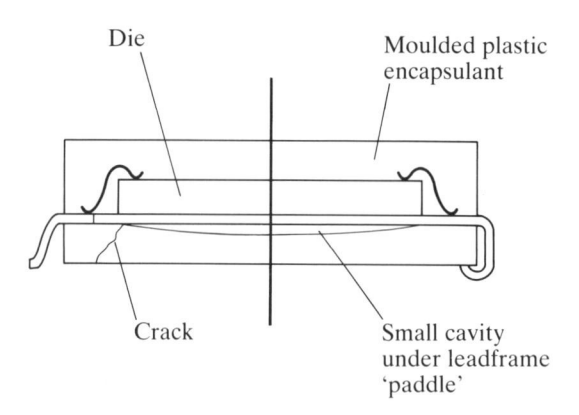

Fig. 17.2 Crack caused by 'popcorn' effect

Packages containing relatively large silicon chips and those with poor lead seals have appeared more prone to the effect. It is also worth noting that many of the larger packages, upon sectioning in their 'as received' state, show a void region immediately below the paddle of the leadframe on which the die sits. This acts as a natural harbour for water vapour.

In addition to gradual moisture ingress through the solid plastic thickness around the device, a major penetration path is along the leads. The joining of the plastic to the metal is only a compression seal, with no chemical bond. It has also been suggested that the change to use of copper-based leadframes has rendered such devices more prone to thermal shock compared with the previous nickel alloys.

The higher thermal conduction and expansion of the copper can then exert considerable stress on the plastic which, on cooling and subsequent temperature cycling during operational life, opens cracks around the leads which allow moisture to creep slowly inwards.

Device manufacturers have a basic problem in selecting plastics for embedding large silicon chips. If a plastic is loaded, for example, with alumina powder to improve its TCE (temperature coefficient of expansion) match with silicon, it is likely to be more susceptible to water absorption. Alternatively, if an unloaded, less permeable material is chosen, then it is unlikely to have a TCE near to that of silicon, which is about 2.7 p.p.m./°C. The consequence of the latter choice is increased risk of damaging the embedded wire bonds due to relative movement of the plastic and the chip during operational temperature cycling.

It must be said that some manufacturers appear better at the art of plastic encapsulation than others, but where there is doubt preventive action is effected in several ways:

1. Manufacturers can be asked to supply their integrated circuit devices sealed in a metallized plastic bag (Drypack) filled with dry gas, or with a small sachet of desiccant. Users are then advised to do their soldering within 24 hours of breaking the seal. Typically, the margin of safety is met if, after breaking the seal, devices are used within 48 hours after storage at 30 °C in air whose relative humidity (RH) is 60 per cent.
2. Just before soldering, users can bake their devices ex-store at, say 45 °C for 200 hours or 80 °C for 48 hours or 100 °C for 24 hours to dry out the moisture. Use of the lower-temperature options is recommended as higher temperatures can cause oxidation giving lead solderability problems—especially in reflow situations.
3. Sample packages can be subjected to rapid heating and examined for cracks before the batch is accepted.

17.4 TCE mismatch

TCE mismatch between component bodies and organic printed board materials is a well-understood source of solder joint fatigue failure. The conditions known to accelerate the phenomenon are:

- Large TCE differentials
- Large distance between solder joints
- Small solder pillar thickness

- Frequent thermal cycling over a wide temperature range below 100 °C
- Increased lead stiffness.

With good design it is practicable to ensure survival over several thousand cycles, depending on the magnitudes of the first four of the above conditions.

A typical descending order of likelihood of causing problems on epoxy–fibreglass printed boards is given in Table 17.1. At the bottom end of the

Table 17.1 Fatigue life: a typical league table

Components[*]	Typical component body materials
1 Ceramic chip carriers	Alumina
2 Large ceramic chip capacitors	Barium titanate
3 Ceramic chip resistors	Alumina
4 MELF components	Alumina/glass
5 Small ceramic chip capacitors	Barium titanate
6 SO and quadpack IC packages	Loaded epoxy
7 SO transistors	Loaded epoxy
8 Plastic leaded chip carriers	Loaded epoxy

[*] 1 represents shortest fatigue life; 8 represents longest fatigue life.

table, the combined effects of lead compliance and body size would be expected to determine fatigue life.

Details of techniques for minimizing the problems are given in both of the author's earlier books. (Refer to the Introduction for details.)

17.5 Other reliability pitfalls

In placing surface mount subcontract assembly business, it is wise to review the board layout with the potential subcontractor's engineers and to tap their knowledge of where things are known in the past to have given field reliability problems. The most important of these are dealt with in Chapters 5, 7, 8 and 10.

17.6 Conclusion

It would be wrong to conclude this chapter without pointing out that, although all the defects described in Sections 17.2, 17.3 and 17.4 are persisting at a low but still unacceptable level, they are occurring less frequently as awareness improves. More people are taking out the relevant 'insurance policies'.

When things do go wrong, however, there has usually been a very heavy consequential cost penalty.

18

Maintaining good relations

Very often the customer–subcontractor interface develops into what might well be described in psychological terms as a 'love–hate' relationship. There is nothing wrong with this: it is the basis of many personal contacts, and it can be helpful for both sides to see it in that light. Each needs the other in order to survive; each respects the other's good points and is angry when the bad ones emerge; but each must live with the frailties of the other—at least until revocation or contract completion.

18.1 Defining respective responsibilities

If this matter has not been dealt with (as it should have been) during the quotation and contract negotiation stages, to avoid arguments on liability it must at least be clarified before delivery commences.

Discussion should centre on the respective design and manufacturing responsibilities and therefore on which of the two or more parties involved is the 'design authority' for each. See Chapter 11 for an example of a typical approach to this subject.

18.2 Departmental contacts

A well-run customer–subcontractor interface requires detailed contact at senior level between respective marketing, design, purchasing, production, and quality managers. It is a common mistake to try to channel all contact through one person on either side.

This will sound like heresy to customers who are inexperienced in handling such an interface involving surface mounted assemblies. They see the need as requiring very close control and are convinced that this is practicable only via a single voice from each party. This might just work if the persons concerned were both very intelligent and highly experienced in the technology. The lessons learned from the past are that the risks of error are often higher through an imperfect single funnel than through a multi-channel link, a phenomenon that arises through the tendency for the person in the funnel position to take apparently straightforward decisions without realizing all their implications.

There is another reason why multiple contacts are important. It is that the subcontractors become, in effect, the equivalent of the major proportion of the in-house production department. As far as information flow is concerned, they should be encouraged to function more like an integral part of the customer's company than as an arm's-length partner.

The more the subcontractor's sales manager knows about the customer's sales plans and problems, the more likely it is that he or she will be motivated to bring strong pressures to bear on the production people. There is often more natural empathy between similar departments in different companies than between different departments in the same company! The same can be true of both quality and production functions. For example, they can share the same exasperations with the design and sales people for landing them with unnecessarily difficult tasks.

Of course, this approach is not possible where either party is a small company or when the subcontractor is a single-executive organization.

18.3 Controlling changes

The customer–subcontractor interface has always required good control of modification procedures. The introduction of surface mounting in this context has brought the need for even tighter disciplines, partly because of the emergence of a far wider range of variables in both design and manufacturing processes and partly because of the greater complexity of board content.

The variables include the interactive chemistry of component, board, solder and cleaning materials, the higher process temperatures seen by components, disciplined footprint geometries coupled with lack of termination standards in the component design and supply chain, and electrical performance changes arising from smaller size and closer proximities of copper track and component bodies, to name but a few.

It is good practice for the customer to ask to examine the subcontractor's Change Note before contract placement. In addition to the usual internal circulation for approvals by production, quality, engineering and sales, a professional document will contain provision for estimating the cost of making the change, including the effect on work in progress and unusable stock, delivery delay information and space for a customer approval signature.

Changes come in two main categories.

(A) MODIFICATIONS INITIATED BY THE CUSTOMER
These amount to changes in the purchasing specification and therefore carry

the implied risk of increase (or advantage of decrease) in the previously agreed price.

After notification is received by the subcontracting firm, control is best exercised by its raising its own Change Note. If the initial data were verbal, then obtaining the customer's signature on the Change Note is essential prior to implementation. This is not only to gain formal approval in writing for the change(s), but also to enable the customer to check that the verbal instructions have been understood and correctly interpreted. By definition, this means that the Change Note must give, or be accompanied by, sufficient information for the customer to make such a judgement.

In the past, subcontractors have often been put under great pressure to proceed without waiting for formal confirmation of changes, but the advent of facsimile (fax) machines has removed most of the delay. Today there is no justification for the customer to use the time factor as a reason for failing to give immediate written commitment. It is worth mentioning that, unlike a telex message, proof of transmission of a facsimile document is not necessarily regarded as legal proof of its receipt in complete and legible form.

(B) CHANGES INITIATED BY A SUBCONTRACTOR

These emerge from a variety of needs, most of which are those that seek to vary the layout design or assembly process to suit the product (or vice versa). If the customer asks to be notified and to approve such changes or the assembler offers these arrangements, effective control must be demonstrable via the subcontractor's Change Note system.

The key decision area centres around what constitutes a 'notifiable' change. Perhaps the best way to handle this situation is for the subcontracting firm to state precisely the changes that it is prepared to notify, in which case it would be expected to keep suitable records of these.

Proponents of Murphy's Law will hasten to point out that it is those 'minor' changes not expected to cause problems that invariably do so—therefore all changes should be notified. Apart from being impracticable in the extreme, this viewpoint is based on the dubious assumption that either or both parties will be capable of forecasting the effects of all the changes, major and minor. If they were able to do so, Murphy's Law would then be invalid—a doubtful proposition!

NOTIFIABLE CHANGES

Suggested guideline notifiable changes by the subcontractor are:

- Change in basic printed board material, e.g. from FR4 to FR2, or in material thickness

- Change in the board copper thickness, layout or hole positions
- Alternative flux type, e.g. change from RMA to RA or vice versa
- Change of solder composition or solder paste metal composition
- Different basic soldering method, e.g. from infra-red reflow to vapour phase reflow
- Major change in cleaning material or process, e.g. from a CFC to a terpene, or from a simple dip wash to ultrasonic cleaning
- Use of an alternative manufacturer for any semiconductor device, multilayer ceramic capacitor, electrolytic capacitor or other specified 'critical' component
- Change in basic rework method for any component, e.g. heated tweezers introduced in addition to, or as a replacement for, a hot gas pencil
- Change of electrical test method for the assembled or part-assembled circuit, including software and/or hardware changes

18.4 Customer acceptance criteria and return procedures for large or complex boards

When a large or complex board is assembled and shipped by the subcontractor solely on the basis of a 100 per cent component in-circuit test, apart from defects that may arise in transit, there is a good chance that it will meet the functional test requirements.

To a large extent, the degree of validity of this statement depends on the circuit complexity and its operating frequency or switching speed, and on the length of time that the board has been in production.

The correct operation of high-frequency circuits depends as much on the tuned interconnections, e.g. the tracks and their impedance parameters, as on the components. The chance of functional success after an in-circuit test is therefore better when the board is one that works at the lower end of the speed–frequency spectrum.

The time-scale factor operates in two ways. In the early phases of production of a surface mounted circuit, it is unlikely that all the components on a complex board will be exercised by the in-circuit test. A typical starting proportion range would be 75–90 per cent, and with the further development of the test software levels in excess of 95 per cent are achieved. Well-known exceptions are application-specific integrated circuits (ASICs), which, apart from being able to prove presence or absence, usually require a functional test for complete verification.

The second time-scale factor is in the electronic design area, where the situation is epitomized by the famous cliche, 'The quality is excellent, but the circuit does not work.'

Aside from partial or complete malfunction due, for example, to logic errors in a new system, the question of marginal parametric deviation arises—especially when a sensitivity analysis on the circuit design has not been carried out. Even if the subcontracting firm has sufficient knowledge of circuit design to diagnose the problem, it can be extremely difficult for it to show that the circuit malfunction stems from poor tolerancing. This is partly because there may be many other variables that are masking the true reason for incorrect performance, and partly because no customer takes kindly to criticism on these grounds.

The important point at issue is whether the testing tasks assigned to the subcontractor and the maturity of the circuit design are together sufficient to justify a goods inwards acceptance based on a board acceptance quality level (AQL). Unless the circuit has been fully debugged and the subcontractor is equipped to do both in-circuit and functional testing, the answer in almost all cases will be 'No'.

What are the reasonable alternatives, and what are their consequences? Acceptance and rejection on an individual circuit-by-circuit basis is the easiest option, but who will locate and define the defects and who will do the necessary rework?

If the customer-firm is carrying out a functional test that has little or no diagnostic capability down to component level, it may or may not be able to pinpoint the fault locations. However, if it has not been able to negotiate provision of an in-circuit test by the subcontractor, then the firm remains fully responsible for identifying the defects and indicating the corrective actions required.

Having found them—and without in-circuit testing this can be a very time-consuming and costly exercise—it may prove cheaper for the customer-firm to do the rework itself rather than circulate the circuits to the subcontracting firm. Apart from the inevitable delays and risk of further damage in transit, there are the additional costs of documentation and transport to consider.

Here we begin part of the argument on 'make or buy' all over again, but the advisability of the 'make' option on rework will be influenced by the level of machine investment and the skill and 'know-how' needed to carry out the rework—which depends greatly on the types of device to be handled. Poor-quality rework is one of the commonest sources of unreliability in surface mounted assembles. Quadpacks with several hundred leads demand a very much greater investment in rework equipment and training than chip passive components and simple 16-lead SO integrated circuits.

Another facet of this situation is that, if the subcontractor does the rework but has no means of deciding whether the circuit is then in good working order, a percentage of second- and third-circulation cycles is inevitable.

The logic from the above situation is that there are two sensible solutions, but that neither is free of problems.

1 The customer-firm accepts untested circuits from the subcontractor and does its own test and rework cycles in-house
In this case, an agreement on the solder joint defect rate in parts per million allowed to the subcontractor is necessary, with some financial penalty if the limit is exceeded, e.g. paying a proportion of the rework costs.

This is satisfactory when the subcontracting firm is in total control of procurement of both the board and the components and cannot then use its poor-quality supplies as a reasonable excuse for exceeding the limit. But when the customer is responsible for free issues and there are problems with their solderability, it can be unreasonable to operate such a penalty system. Much will depend on who has done the goods inwards inspection and whether the critical items have been supplied in time for solderability tests to be carried out.

2 The subcontracting firm is equipped with both in-circuit and functional test equipment and all components are exercised by it
In this instance, AQLs at goods inwards are considered reasonable once the circuit has been debugged. Prior to that event, acceptance on a board-by-board basis is the correct approach. Regular cross-checking of the functional testers held at the subcontractor's and customer's premises is essential.

In both of the above scenarios there is a presumption of acceptable 'design for manufacture' which, if not fulfilled, can lead to argument if things go awry. Ideally, any foreseeable assembly difficulty on these grounds should have been reflected in the price charged by the subcontractor and it is good policy to try, as far as possible, to take this factor into account when setting the agreed p.p.m. defect level in the first place.

Having said this, when there are several hundreds of components on a board there is always a possibility that some design lapses may not be spotted before production starts—in which case one may expect the subcontractor to come knocking on the buyer's office door, 'brickbat in hand' or otherwise!

In reviewing options 1 and 2 above, it must be said that intermediate solutions are practicable if the circuit is not complex, if the pincount is not high (e.g. <250) and if the locations of the subcontractor and customer are only a few miles apart.

One local arrangement that can work, even if the board is complex, is that the subcontractor provides and manages a rework facility on the customer's site for the period of the contract. This can be a good solution because it enables maximum advantage from both skills: the subcontractor's trained rework staff and the customer's test equipment and circuit expertise. Properly constructed, the arrangement can also keep assembly quality high because the subcontracting firm will seek to minimize its commitment to an 'away game'.

18.5 Product safety and product liability issues

In some countries, for example the UK and the USA, the designer of an unsafe equipment that causes harm (or the managing director) may risk going to gaol. But in the product safety context, a responsible and professional purchasing function may also end up in trouble if it has failed to look at both incoming material and outgoing product responsibilities.

In many countries, the law now requires that all products sold to industry must have user product safety information available, and most of this will enter the company through the purchasing department. There is therefore a need, first, to ensure that these data reach those who are at risk and, second, where it is not supplied—to demand it. When an assembly is to be subcontracted, this responsibility must be extended to cover the subcontractor—especially if there are free-issue materials involved.

There is also a requirement to ensure that all materials and equipment supplied to or specified to a subcontracting firm are 'fit for purpose'. In this respect they must enable the firm to fulfil its contract provided it exercises due care and deploys the tools and skills of its trade in a reasonable and diligent manner.

While there may be plenty of scope for argument when problems arise, the items mentioned above are regarded as routine procedures which should be followed in implementing both the law and common-sense actions.

In product liability matters, things are more complex. The law in European Community countries has been changed in favour of the general public in that, if a person is harmed by a product, it is no longer necessary for him or her to establish negligence on the part of the seller or manufacturer. With today's levels of technology in most industries, this was extremely difficult for non-technical and semi-technical legal eagles to prove. Now, a complainant is required to show only that the product caused the harm and that the product was marketed by the defendant.

A second problem was that the complainant was able to sue only the person or organization that sold him or her the product, as this relationship was considered to contain the relevant contractual responsibilities. Under the new 'strict liability' rules, they can pursue the original source of the problem directly.

The implications of these points of law materially affect the subcontract scenario, and normally it falls to the purchasing function to acquire the necessary legal support and to insert prudent terms into contractual documents. In particular, it should be appreciated that typical 'Terms and Conditions of Purchase' and 'Terms and Conditions of Sale' documents are unlikely to contain the required texts. Some suggested approaches to this problem are given in Chapter 3, though it should be noted that words in contractual documents may not give total protection in the event of a mishap involving professional incompetence or negligence. In any event, proper legal advice should be obtained.

Where there are identified risks in the product, a first step is to establish whether the intended subcontractor has product liability insurance cover. It is important to establish its magnitude, the name of the insurers, and whether it applies to the countries in which the final product is to be marketed. Purchasing managers in large companies often prefer to deal only with subcontractors who have reasonable cover.

Conversely, sometimes the subcontracting firm will ask for a document confirming that it will be 'held harmless' in the event of a product liability attack. Such a document may offer temporary protection but is no longer a complete defence against proceedings being taken if culpability is clear; cf. Appendix 3A, paragraph 8.

Unfortunately, in countries where the premiums are excessive, many subcontractors prefer to avoid the cost and 'go bare' with no insurance and the implicit acceptance that, if there is a major problem, they will probably be bankrupted.

As previously indicated, both parties should have an interest in establishing in writing their respective responsibilities in the design, manufacture and test sequence. The professional subcontracting firm will attempt this in its quotation document.

It is worth mentioning that, while most product liability insurance policies cover the cost of recall, damages and legal costs, usually they do not extend to the cost of redesign and replacement of defective parts. Both of these can be capable of bringing a call to receivership.

19
Costing board assemblies

19.1 Background comments

An understanding of the methods of cost estimating and cost collection used by a subcontractor can be helpful when comparing quotations.

The team that is responsible for preparation of the cost estimates to be used in quotations is not always located within the same departmental function. Professional subcontractors offering a board layout service will probably assign the task to their engineering design group and use a sophisticated spreadsheet method based on a mixture of standard and/or synthetic operation times. Those basing their quotation system on machine times are more likely to give the job to the production department.

The ongoing collection of actual production cost data on a circuit-by-circuit basis is certain to be more complex for a subcontractor than for an OEM and many smaller operations do not bother with it unless they have a long-running project. A large part of the problem arises because the work prior to soldering and cleaning is handled in multiples, i.e. identical circuits in step-and-repeat panels, and then after break-out it converts to individual circuits. Unless great care is exercised, the system can become top-heavy with paperwork.

In situations where some of the shop-floor workforce are trained to have multiple skills and others are not, usually it is cheaper and more reliable to collect time-booked data from operators' daily record cards than from travelling batch documents, though in this case the cards must include space to fill in time spent on individual job and batch numbers.

Another important factor in a subcontractor's facility is that waiting or idle time percentages tend to be higher. This is due as often to customers' free-issue delays and other procurement difficulties as to work planning deficiencies. In this context, pick-and-place machine operating problems can be very disruptive.

All cost-collection systems are dependent on the integrity of workforce members when filling in their daily records. The absence of piece-rate pay methods helps in this context, but other means of stimulating good output are essential.

Some differences in costs and costing have emerged in the industry when it comes to working on surface mounted assemblies as compared with earlier technologies. These are indicated below.

PRODUCTION YIELD LOSSES

Allowances for component yield loss have to be larger, especially in the early stages of product development, because complete circuits may have to be written off while optimizing design and process parameters. The underlying factors here are the greater number of components per board and the higher temperature seen by them during soldering, e.g. 250 °C instead of 120 °C.

Worth noting too is the ability of pick-and-place machines to launch components into the air at a very fast rate if they, or the pick-up or feeder systems that handle them, are slightly out of kilter.

STEP-AND-REPEAT ARRAYS

To reduce manufacturing costs, most boards are carried in step-and-repeat panel arrays whose content may range between 2 and 100 identical circuits per panel. In some instances the panels remain whole from paste or adhesive deposition right through to electrical test, but in most cases break-out into individual circuits occurs before testing.

REWORK

The increased difficulty of reliable rework means that this activity forms a higher proportion of the factory prime cost. Part of this increase results from the fact that most of the assembly is carried out by mechanized processes at low cost, whereas the rework remains a manual and/or component-by-component operation. The related problems and precautions are discussed in Chapter 16.

CAD FOR CIRCUITS AND LAYOUTS

Electronic circuit design, printed board layout and associated software and hardware are all higher in cost per board, but are often comparable on a 'per pin' basis.

The need to get the electronic circuit design right first time has increased the importance of affording computer programs for design checking, e.g. schematic capture, sensitivity analysis and tolerancing.

PRINTED BOARDS

Purchased bare board price 'per unit area' has increased by factors ranging from ×2 to ×10 and per component from ×1.5 to ×2.5. However, the price 'per pin' has not risen and is sometimes less than for conventional through-hole boards.

LEARNING CURVE

Currently the learning curve costs are higher on the assembly line for each new circuit design using surface mount technology, mainly because of the increase in component packing density and board complexity. This fact is of even greater importance to subcontractors who have to cope with a very high rate of new product introduction compared with OEMs operating in-house manufacture.

19.2 Labour-based cost-estimating methods

For subcontractors, labour-time-based methods of cost estimating remain the most accurate option, principally because they can be compared more easily with later reality by using time-booking procedures. This form of feedback helps to generate the standards and the related difficulty factors that are applied. These factors are used to take account of variations, for example in complexity and/or in component packing density. However, they take longer to prepare, and those companies specializing in very fast turn-round business with a wide range of circuits on their assembly line at any one time are not likely to collect operator timesheets giving data on individual circuit types.

On high-volume production lines, the proportion of labour in the prime cost can be very small—perhaps less than 10 per cent—and the value in collecting operator-time information is correspondingly lower.

Another major advantage of labour-based methods is that they can be applied to a wider range of operations, particularly those prior to reaching the production line and after leaving it. The time spent on kitting in the stores is highly product-sensitive, as are packing and shipping. In the past these items would have been consigned to the overheads, but in today's competitive times they have become direct labour costs alongside many of the quality control and production engineering tasks.

Although the creation of synthetic operator throughput rates demands work study, the data collection is required only for estimating and not for determining piecework rates. Hence the difficulties that beset work study engineers in trying to agree standard times with trade union representatives are absent.

A compressed example of a typical labour-based cost estimate is given in Tables 19.1 and 19.2.*

19.3 Machine-based cost estimating methods

In this system, each process group in the assembly facility is assigned a minute rate based on the annual cost of operating it—assuming that it is in continuous use for the relevant shift time. For example, the groups may comprise solder paste printing, auto-placement, infra-red soldering, ultrasonic cleaning and electrical test.

To estimate the production prime cost of a new circuit, the machine usage times and costs forecast for each operation are compiled and added together to give the total cost of passing the circuit through the designed process sequence. These costs, with allowances for set-up, machine downtime and yield losses, become the 'standards'. Excess yield loss and under-utilization of the machines will show up as variances.

If the method is the only one used to obtain the total prime cost, then any manual operations have to be linked to a relevant machine and become a part of the running costs associated with it. For example, the salary of the operator who runs a pick-and-place machine is added to the annualized depreciation and other operating costs for that machine, enabling an average hourly rate to be calculated.

For example, a pick-and-place machine rated at 15 000 component placements per hour might cost £40–50 per hour to run.

19.4 'Scarce resource' costing

A 'scarce resource' is defined as that facility which limits production by imposing an output bottleneck.

In this method, forecast fixed production costs for the entire assembly line for the year—including overheads—are used to define an hourly rate for the nominated 'scarce resource', say £A. The scarcity can take many forms; for example, it may come from a materials supply difficulty, a shortage of trained labour or a machine throughput limitation.

---

* The spreadsheets shown in Tables 19.1–19.4 are part of a linked set of five applying labour-based cost-estimating methods. They use Lotus 123/2.2 software and are available on diskette with a user instructions leaflet from the Surface Mount Club, Building 15, National Physical Laboratory, Queens Road, Teddington, Middlesex, TW11 0LW (tel. 081-943-7150). The spreadsheets are intended for use by CAD designers and others who carry out board layout or estimating. Apart from providing a Bill of Materials on Sheet 1 (Table 19.1), the fifth sheet (not shown in this book) contains dimensions of most available surface mounted components, and these data are used to calculate the percentage board occupancy figures needed to determine likely copper layer count. They also enable total assembly volume estimates.

Table 19.1 Surface mount and mixed technology component and board cost and size estimate/bill of materials

Components for Board No...... ...of......... Circuit Type No.......... Project:.......... Assembly Type : Side 1 TH + SM / Side 2 SM only
Drawing No :.......... Issue No.......... Date of Estimate:.......... Solder Method : Side 1 Reflow, 1 pass / Side 2 Wave, 1 pass
Max Cct Board Dimms : 100.0 40.0 1.6 mm Circuits/panel 8.000 (from COSTPROC File)
Max Cct Assy Volume : 6.4 cc → G H
PCB Panel Size/Type : 300 × 250, 6-layer, FR4

PCB & Compt Types	Format of Compt	Supplier	Supplier's Ref No	Compt Qty/Cct	No of Pins/Compt	Compt Qty P&P	Compt Qty Insert	Side 1 Unit Pad Areas mm²	Max Side 1 Mntd Compt Hight 6.0 mm	Side 2 Unit Pad Areas mm²	Max Side 2 Mntd Compt Hight 3.2 mm	Production £ Compt Cost Each	Production % Est Compt yield	Total Purchase Order Qty 100,000	Total Compt Costs /Cct	Production £ Compt Cost Each	Production % Est Yield	Total Purchase Order Qty 500,000	Total Compt Costs Total	File: Company's Part No	CostComp. Wk1 NOTES
Printed Circuit	Panel, st≥p ≥ repeat	Pushwick Ccts		1								64.00	98.0	102,041	8.163	54.00	99.0	505,051	6.818		
Processor No 1	VSO 40 package	Plodson		1	40	1		200	3.0			9.800	99.0	101,010	9.899	7.500	99.0	505,051	7.576		
A/D converter	DIL	Carroller		2	24	1	2	165	6.0			3.400	99.5	201,005	6.834	2.750	99.5	1,005,025	5.528		
Modulator	Quadpack (→SIC)	Eastelec		1	64	3		375	3.0			5.850	99.0	101,010	5.909	4.900	99.0	505,051	4.949		
Ceramic capacitor	Chip 1206 42nF 50V	Rapcata		3	2	3				10	2.2	0.040	98.0	306,122	0.122	0.032	98.0	1,530,612	0.098		
Resistor	Chip 1210 1E 2%	Agetoile		2	2	2				13	1.2	0.035	97.5	205,128	0.072	0.031	98.0	1,020,408	0.063		
Resistor	Chip 0805 22K 5%	Agetoile		2	2	2				7	1.2	0.017	97.0	206,186	0.035	0.015	97.5	1,025,641	0.031		
															0				0.000		
Processor No 2	VSO 40 package	Plodson		1	40	1		200	3.0			4.100	99.0	101,010	4.141	3.650	99.0	505,051	3.687		
ROM 64K×8	DIL	Tego		1	28	1	1	390	6.0			3.500	99.0	101,010	3.535	3.050	99.0	505,051	3.081		
RAM 16K×8	DIL	Tego		1	28	1	1	200	6.0			5.500	99.0	101,010	5.556	4.500	99.0	505,051	4.545		
Processor No 3	VSO 40 package	Plodson		1	40	1		330	3.0			2.950	99.0	101,010	2.980	2.550	99.0	502,513	2.576		
RAM 2K×8	DIL	Tego		1	24	1	1	200	6.0			1.600	99.5	100,503	1.608	1.400	99.0	1,025,641	1.437		
Multichip	VSO 40 package	Plodson		2	40	2	2	200	5.0			2.450	97.0	206,186	5.052	2.200	97.5	1,025,641	4.513		
Crystal	6X6×14.5 × 5 × 12	Fifty Int'l		1	2	1				32	1.8	0.790	99.5	201,005	1.588	0.750	99.5	1,005,025	1.508		
Ceramic capacitor	Chip 222 0.1 mF 50V	Simonron		1	2	1				10	2.5	0.320	98.0	102,041	0.327	0.275	98.0	510,204	0.281		
Ceramic capacitor	Chip 1206 10nF 50V	Simonron		1	2	1				10	2.5	0.045	99.0	202,020	0.091	0.040	99.0	1,010,101	0.081		
Resistor	Chip 1206 47K 10%	Agetoile		1	2	1				10	1.2	0.015	97.0	103,093	0.031	0.013	97.0	515,464	0.013		
Resistor	Chip 1206 1.5K 5%	Agetoile		2	2	2				10	1.2	0.017	97.0	206,186	0.035	0.015	97.0	1,030,928	0.031		
															0				0.000		
Channel codec	VSO 40 package (ASIC)	Plodson		1	40	1		200	3.0			6.100	99.0	101,010	6.162	5.850	99.0	505,051	5.909		
Ceramic capacitor	Chip 1206 1.0nF 50V	Simonron		2	2	2				10	2.5	0.062	97.0	206,186	0.128	0.075	97.5	1,025,641	0.154		
Ceramic capacitor	Chip 1206 15nF 50V	Simonron		1	2	1				28	2.8	0.075	97.0	206,186	0.155	0.069	97.5	1,025,641	0.142		
Ceramic capacitor	Chip 1812 68.8nF 50V	Simonron		1	2	1				23	2.0	0.360	97.0	103,093	1.371	0.031	97.5	512,82	0.132		
															0.000				0.000		
PCM codec	SO 28L package	Hitrun		1	28	1		200	3.0			3.950	99.0	101,010	3.990	3.600	99.0	505,051	3.636		
Transcoder	VSO 40 package (ASIC)	Plodson		1	40	1				200	3.0	6.250	99.0	101,010	6.313	5.900	99.0	505,051	5.960		
Resistor	Chip 1206 10K 10%	Agetoile		2	2	2				10	1.6	0.015	97.0	206,186	0.031	0.013	97.0	1,030,928	0.027		
Resistor	Chip 1206 1K5 10%	Agetoile		2	2	2				10	1.6	0.015	97.0	103,093	0.015	0.013	97.0	515,464	0.013		
Resistor	Chip 1206 3K 10%	Agetoile		2	2	2				10	1.6	0.015	97.0	103,093	0.015	0.013	97.0	515,464	0.013		
Resistor	Chip 1206 22K 10%	Agetoile		1	2	1				10	1.6	0.015	97.0	103,093	0.015	0.013	97.0	515,464	0.013		
Resistor	Chip 1206 1K5 10%	Agetoile		1	2	1				10	1.6	0.015	97.0	103,093	0.015	0.013	97.0	505,051	0.013		
Electrolytic cap	TAT 4.7mF13V 20%	Paintex		2	2	2				22	3.2	0.340	99.0	101,010	0.343	0.290	99.0	505,051	0.293		
Electrolytic cap	TAT 6.8nF 4V 20%	Paintex		1	2	1				22	3.2	0.350	99.0	101,010	0.354	0.300	99.0	505,051	0.303		
															0.000				0.000		
>															0				0.000		
To 3000 Compts max															0				0.000		
>															0				0.000		
>															0				0.000		
etc															0				0.000		
															0				0.000		
		TOTALS		42	556	35	7	3615		555		Compt	cost per Cct :	101,010	73.87	Compt	cost per Cct :	505,051	63.32		
				83	*Total pincount			90		14											

Component area as % circuit are :
Quality Factor : 0.13× H120 + 0.25 ≤ 5120 :
Output rate Max Ccts per week 5000 /Wk 7500 Wk
Value of Free Issue Stock per Circuit 66.0 56.5
Subcontract only Free Issue Est value of 1 week stock £ 330000 £ 423750
Subcontract only Free Issue Insce Prem @ ≈ .060 % per week £ 198 £ 254 Max

Table 19.2 Surface mount and mixed technology printed board assembly cost estimate

ASSY PROCESSES FOR BOARD No.... of ....   CIRCUIT TYPE No ....   Date of Estimate ....

DRAWING NO .... Issue No. ....   Panel Batch Sizes ———————→

Loaded Labour Rate. Assy   £0.300   <——
Loaded Labour Rate. Test   £0.400   <——
No of Circuits/Panel   8   <——

SM&TH Compt Qty's/Cct		Op'n % or Sample %	Panels per Op'n	Ccts per Op'n	Set-up Mins /Batch	Mins per Op'n	2 Protos Op'n Mins /Cct	2 Protos Set-up Mins /Cct	120 Prodn A Op'n Mins /Cct	120 Prodn A Set-up Mins /Cct	600 Prodn B Op'n Mins /Cct	600 Prodn B Set-up Mins /Cct	Inspn Secs/ Joint	Defect level ppm	NOTES
	Operations														
	Kitting		125	1000	10	60.00	0.60	0.625	00.060	0.010	0.060	0.002			
	Clean Printed board	100	5	40	5	5.00	0.125	0.313	0.125	0.005	0.125	0.001			
750	Solder paste print	100	1	8	20	1.00	0.125	1.250	0.125	0.167	0.125	0.033			
	Auto-placement	100	1	8	300	22.50	2.813	18.750	2.813	0.313	2.813	0.063			
	(M/c rate per hr) →	16000	Components (2 M/cs)												
	Reflow solder	100	5	8	120	2.00	0.250	7.500	0.250	0.125	0.250	0.025			
	Clean	100	5	40	5	5.00	0.125	0.313	0.125	0.005	0.125	0.001			
	(No of passes)	1	(Op'ns 15, 16 & 18)												
100	Glue spot & cure	100	1	8	30	1.00	0.125	1.875	0.125	0.031	0.125	0.006			
	Auto-placement	100	1	8	100	10.67	1.333	6.250	1.333	0.104	1.333	0.021			
	(M/c rate per hr →	4500	Components												
50	Hand insertion	100	1	8	15	20.00	2.500	0.938	2.500	0.016	2.500	0.003			
	(Hand rate per hr) →	150													
120	Auto-insertion	100	1	8	120	11.52	1.440		1.440	0.125	1.440	0.025			
	(M/c rate per hr) →	5000	Components												
	Wave solder	100	1	8	10	3.00	0.375	0.625	0.375	0.010	0.375	0.002			
	Clean	100	5	40	5	5.00	0.125	0.313	0.125	0.005	0.125	0.001			
	(No of passes)	1	(Op'ns 28 & 30)												
	Break-out	100	1	8	10	1.00	0.125	0.625	0.125	0.010	0.125	0.002			
	Special Op'ns 'A'	100	1				0.000	0.000	0.000	0.000	0.000	0.000			
	Special Op'ns 'B'	100	1				0.000	0.000	0.000	0.000	0.000	0.000			
	Special Op'ns 'C'	100	1				0.000	0.000	0.000	0.000	0.000	0.000			
	Special Op'ns 'D'	100	1		10	5.00	0.500	0.313	0.500	0.005	0.500	0.001			
	Special Op' Clean	100	1	10	5	5.00	0.500	0.313	0.500	0.005	0.500	0.001			
	Visual Insp'n No 1	100%	of boards	1	20	5	41.67	1.250	41.67	0.021	41.67	0.004	1.0		
	Rework Op'n No 1	100%	of boards	1	5	5	10.00	0.313	10.00	0.005	10.00	0.001	120.0	2000	2000ppm. 2min/joint
	Clean	100%	of boards	10	20	5	0.500	0.313	0.500	0.005	0.500	0.001			
	In-Cct El Test No 1	100%	of boards	1	20	5	3.000	1.250	3.000	0.021	3.000	0.004			
	Rework Op'n No 2	50%	of boards	1	5	5	1.000	0.625	1.000	0.010	1.000	0.002			
	Visual Insp'n No 2	50%	of boards	1	5	5	0.083	2.500	0.083	0.042	0.083	0.008	120.0	200	200ppm. 2min/joint
	In Cct El Test No 2	50%	of boards	1	20	5	1.500	1.875	1.500	0.031	1.500	0.006	10.0	200	
	Functional El Test	100%	of boards	1	30	5	5.000	25.000	5.000	0.417	5.000	0.083			
	In Cct El Test No 3	5%	of boards	1	20	3	0.150	6.250	0.150	0.104	0.150	0.021			
	Rework Op'n No 3	5%	of boards	1	5		0.150	25.00	0.150	0.417	0.100	0.083			
	Functional El Test	100%	of boards	1	20	5.00	4.167	6.250	4.167	0.104	4.167	0.083	120.0	20	20ppm. 2min/joint
	Final Visual Insp'n	100%	of boards	1	5	0.50	0.500	0.313	0.500	0.021	1.250	0.001		20	
	Marking	100%	of boards	10	20	3.00	4.167	1.250	4.167	0.021	4.167	0.004	0.1		
	Packing	100%	of boards	10	10	3.00	0.300	0.625	0.300	0.010	0.300	0.002			
1020	← Total SM+TH Compt's/Cct														
2500	← Pins=Joint Count (From Component sheet)	Total Mins per Circuit →					76.80	106.88	78.24	2.05	78.24	0.41			
	Total Set-up time & Cost/Cct						Mins	£	Mins	£	Mins	£			
	Op'ns Time & Cost/Cct (Excluding Set-up)						534.38	165.88	2.67	0.89	0.41	0.14			
	Op'ns Learning Curve Multiplier						383.99	116.19	101.71	31.50	78.24	24.46			
							5.0		1.3		1.0				
	Total Op'ns + Set-up. Time & Labour		——→ cost/Cct Clean				Mins	£	Mins	£	Mins	£			
			——— No Clean				918.4	282.1	104.4	32.4	78.6	24.6			
									103.1	30.9	77.4	23.2			
Memo	Total Rework. Time & Cost/Cct						55.5	16.7	14.4	4.3	11.1	3.3			

In surface mounting, the pick-and-place machine is usually selected for this calculation. The rate is set by determining the total maximum annual working hours of the equipment, including all downtime and setting-up time, say, B hours. The hourly rate is then £(A/B) per hour.

If the effective placement rate of the machine is C components per hour, a circuit requiring D placements will be costed at £$(D/C) \times (A/B)$. Currently the range of high-volume rates seen in the UK market-place varies between £90 and £140 per hour for a professional subcontractor in medium-scale volume (1000–5000 circuits/week) and from £80 to £100 per hour for very high-volume assembly of a simple circuit.

The advantages of this method are its simplicity and ease of use, leading to rapid response quotations—a big marketing 'plus' point.

However, the method is inaccurate and unsuitable where additional hand assembly is significant, e.g. for over 5 per cent of placements. Also, subsequent comparison of the actual production costs with the original estimate can be difficult to determine on multi-product lines.

19.5 Mixed machine and labour-based costing

This option is probably the most accurate, especially when there is high proportion of labour-intensive operations—but it is also likely to be most expensive to run.

The total workload is partitioned into activity areas in which the costings are either machine-based or labour-based. Costs are estimated by adding the respective times in each area and calculating the proportions of the relevant hourly rates carried by each circuit handled.

The methods of arriving at the area rates are identical to those used in Sections 19.2 and 19.4 above.

19.6 Typical labour-based subcontractor cost estimate

GENERAL POINTS

The cost estimate shown in Tables 19.1 and 19.2 is constructed in two parts, first for direct materials and second for the process operations. A third part would be added to cover multiple-board assemblies as in Table 19.3 and, in the case of system design and tooling, a fourth part is needed, typically as in Table 19.4.

To get from the estimate to the quoted price involves setting handling and inventory holding charges for the materials and, in total, achieving an added value to give the required percentage of selling price for profitability.

For example, where V = added value, S = selling price and M = yielded materials, then $V = (S - M)$; and for a typical specialist professional surface mount assembly subcontractor $(S - M)/S \times 100 = 55$–$60\%$, depending on project magnitude.

Table 19.3 Equipment assembly and test operations cost estimate

PROJECT No .... PROJECT NAME .... DRAWING No ....

DATE OF ESTIMATE ....

£ 0.30 Loaded Labour Rate – Assy
£ 0.40 Loaded Labour Rate – Test
£ 0.35 Loaded Labour Rate – Average

ITEMS	QTY	SUPPLIER	MANUFACTURERS PART No	EQUIPMENT ASSEMBLY OPERATIONS	Joint Count	Board Assy Labour £	Matls Each £	Matls Total £	Op'n Mins /Cct	Set-up Mins /Batch	Board Assy Labour	Matls Each	Matls Total	Op'n Mins /Cct	Set-up Mins /Batch	Company Part No	ppm Defect Rate	Rewk Mins/ Joint	NOTES
Flexibles kit	1	Kite Ltd	FL 40765		122		2.75	2.75				2.60	2.60	1.9	10	FB 7123956A			
Board No 1	1	Factory B	FB 77771	Connect flexibles		4.69	32.40	32.40	2.0	10	4.45	31.23	31.23	2.4	15				
Board No 2	1	Factory B	FB 77772	Connect flexibles		3.97	64.23	64.23	2.5	15	3.81	62.21	62.21	2.8	12				
Board No 3	1	Factory B	FB 77773	Connect flexibles		3.60	68.02	68.02	2.0	15	3.50	60.50	60.50	9.0	20				
Board No 4	1	Factory B	FB 77774	Connect flexibles		3.42	44.97	44.97	10.0	20	3.38	42.10	42.10	0.0	0				
Boxbase (Unit front)	1	Ace mouldings	92/564821				0.65	0.65	0.0	0		0.60	0.60	1.0	20				
Metallic linings kit	1	BB Electronics	SMY 4326	Fit to boxbase	6		6.00	6.00	1.0	20		5.00	5.00	1.0	10				
Tone Caller	1	CD Plastics	LSMIN 64/1	Mount caller to boxbase	2		0.45	0.45	1.0	10		0.39	0.39	1.0	10				
Antenna	1	Aerialec ltd	7755312	Mount antenna to boxbase	1		2.09	2.09	0.6	20		1.99	1.99	0.6	20				
Microphone	1	BB Electronics	MICMIN 41/2	Mount mike to Board 2	4		0.65	0.65	1.4	15		0.55	0.55	1.4	15				
Earpiece	1	BB Electronics	PHONMIN 2345	Mount earpiece to Board 1	4		0.75	0.75	1.5	10		0.67	0.67	1.3	10				
Antenna lead	1	CabelCo Ltd	33SH56	Connect antenna lead	2		0.40	0.40	1.5	15		0.39	0.39	0.9	15				
				Functional test 100%				0.00	10.0	60			0.00	8.0	60				
				Rework 3000ppm .5min/jnt				0.00	2.1	10			0.00	2.1	10		3000	5.0	Board defect %
				Functional test 43%				0.00	4.2	30			0.00	4.2	30				42.3
				Rework 300ppm. 7mins/jnt				0.00	0.3	0			0.00	0.3	0		300	7.0	
Board assembly				Assembly boards to boxbase				0.00	5.0	15			0.00	4.5	15				
Battery case	1	EFG Mouldings	90003	Insert case to boxbase			0.30	0.30	1.0	10		0.28	0.28	1.0	10				
Battery connector kit	1	Kite ltd	BC 59022	Solder wires to Board 3	2		0.25	0.25	1.5	10		0.22	0.22	1.4	10				
Boxlid (Unit back)	1	Ace Mouldings	92/564822	Insert lid to boxbase			0.35	0.35	0.5	10		0.31	0.31	0.5	10				
6BA screw	1	FGH Parts Ltd	S4579034	Screw boxlid to boxbase			0.01	0.01	0.5	5		0.01	0.01	0.5	5				
				Functional test 100%				0.00	10.0	0			0.00	8.0	0				0.4
				Rework 30ppm. 10mins/jnt				0.00	0.1	0			0.00	0.1	0		30	10.0	
				Functional test				0.00	1.5	60			0.00	2.5	60				
				Soak test 4 hrs				0.00	3.0	0			0.00	8.0	0				
				Final functional test				0.00	10.0	0			0.00	0.2	0				0.5
				Trouble-shooting/rework				0.00	0.3	0			0.00						
Battery charger	1	Power Supplies	84-96-76				5.20	5.20				5.01	5.01						
								0.00					0.00						

Totals Mins :

Totals £/Equipment £ 229.47 229.47 72.7 360.0 214.06 214.06 63.7 357.0

Labour + OH £ 25.7
Materials £ 229.47
TOTAL 255.17

Total prime cost/Equipt £ 22.44
214.06
236.50

230

Table 19.4 Surface mount and mixed technology board and system development cost estimate

£ 32.10 Hourly Rate : Devt
£ 25.50 Hourly Rate : Devt Assy

PROJECT NAME/REF No Armageddon Mark 7 DEVELOPMENT ITEMS	<--- SYSTEM --->			<--- BOARD No 1 --->				<--- BOARD No 2 --->				<--- BOARD No 3 --->				<--- BOARD No 4 --->				£ Total Devt Cost
	In-Hse Man Wks	£ In-Hse Labour Cost	£ Ext Purch Cost	In-Hse Man Wks	£ In-Hse Labour Cost	£ Ext Purch Cost		In-Hse Man Wks	£ In-Hse Labour Cost	£ Ext Purch Cost		In-Hse Man Wks	£ In-Hse Labour Cost	£ Ext Purch Cost		In-Hse Man Wks	£ In-Hse Labour Cost	£ Ext Purch Cost		
DESIGN																				
System Design	65.0	83 460	0	0.0	0	0		0.0	0	0		0.0	0	0		0.0	0	0		83 460
Electronic circuit design	30.0	38 520	0	0.0	0	0		0.0	0	0		0.0	0	0		0.0	0	0		38 520
Breadboard & trials	26.0	33 384	5500	0.0	0	0		0.0	0	0		0.0	0	0		0.0	0	0		38 884
Redesign	13.0	16 692	0	0.0	0	0		0.0	0	0		0.0	0	0		0.0	0	0		16 692
Sensitivity analysis	4.0	5136	0	0.0	0	0		0.0	0	0		0.0	0	0		0.0	0	0		5136
Tolerancing	2.0	2568	0	0.0	0	0		0.0	0	0		0.0	0	0		0.0	0	0		2568
Schematic capture	1.0	1284	0	0.0	0	0		0.0	0	0		0.0	0	0		0.0	0	0		1284
Structural design	6.0	7704	2200	0.0	0	0		0.0	0	0		0.0	0	0		0.0	0	0		9904
CAD layouts	8.0	10 272	0	0.0	0	0		0.0	0	0		0.0	0	0		0.0	0	0		10 272
Thermal analysis	4.0	5136	0	0.0	0	0		0.0	0	0		0.0	0	0		0.0	0	0		5136
Other items	10.0	10 200	0	0.0	0	0		0.0	0	0		0.0	0	0		0.0	0	0		10 200
PROTOTYPES																				
CAD relayouts				0.6	770.4	0		0.4	513.6	0		0.5	642	0		3.0	3852	0		5778
Photoplot				0.0	0	180		0.0	0	120		0.0	0	160		0.0	0	0		460
Step & repeat photography				0.0	0	50		0.0	0	50		0.0	0	50		0.0	0	0		150
Drill & rout disc				0.0	0	35		0.0	0	35		0.0	0	35		0.0	0	0		105
Courier etc.				0.0	0	100		0.0	0	100		0.0	0	100		0.0	0	0		300
Bare board test hardware				0.0	0	1750		0.0	0	1100		0.0	0	1400		0.0	0	0		4250
Bare board test software				1.0	1284	1450		0.6	770.4	1000		0.0	0	1200		0.0	0	0		5704
Solder print screens/stencils				0.0	0	400		0.0	0	0		0.0	0	0		0.0	0	0		400
Printer tooling				0.0	0	0		0.0	0	0		0.0	0	0		0.0	0	0		0
Pick & place off-line program				0.0	0	0		0.0	0	0		0.0	0	0		0.0	0	0		0
Pick & place tooling				0.0	0	0		0.0	0	0		0.0	0	0		0.0	0	0		0
Adhesive depn offline program				0.0	0	0		0.0	0	0		0.0	0	0		0.0	0	0		0
Adhesive deposition tooling				0.0	0	2700		0.0	0	1600		0.0	0	2400		0.0	0	3500		6700
In-circuit test hardware				0.0	0	3000		0.0	0	1950		0.0	0	2350		0.0	0	3250		7300
In-circuit test software				0.0	0	150		0.0	0	100		0.0	0	100		0.0	0	150		350
Soldering operations tooling				0.0	0	500		0.0	0	0		0.0	0	0		0.0	0	0		500
Other development tooling				0.0	0	2500		0.0	0	1000		0.0	0	1250		0.0	0	2700		8014
Type Approval testing				1.0	1020	0		0.6	612	0		0.6	612	0		1.0	1020	0		9630
Redesign				2.0	2568	0		1.5	1926	0		1.0	1284	0		3.0	3852	0		
Other items				0.0	0	1000		0.0	0	500		0.0	0	500		0.0	0	750		2000
Total In-house Man-Weeks Total Development costs	169.0	214 356	7700	4.6	2054.4	10 315		3.1	3906	6055		2.1	642	7795		7	3852	11 055		186 254 053
OTHER OUTPUT DATA	ManWks	£	£	ManWks	£	£		ManWks	£	£		ManWks	£	£		ManWks	£	£		£
Net list	0.0	0	0.0	0.6	770.4	0		0.5	642	0		0.6	770.4	0		1.5	1926	0		4109
Bill of materials	4.0	4080	0.0	1.0	1020	0		0.7	714	0		0.8	816	0		2.0	2040	0		8670
Assembly drawings	8.0	8160	500.0	2.0	2040	300		1.5	1530	500		1.8	1836	1000		2.5	2550	1500		18 416
Other items	0.0	0	0.0	0.0	0	0		0.0	0	0		0.0	0	0		0.0	0	0		0
	4.0	4080		1.0	1020			1.0	1020			1.0	1020			2.0	2040			9180
Total In-house Man-Weeks Total Other output data	12.0	16320	500	3.6	4850.4	300		2.7	3906	500		3.2	4442.4	1000		6.0	8556	1500		28 40 375
Grand Total All Man-Weeks Grand Total All Extl Purchase Grand Total All Costs	181	230 676	8200	8.2	6904.8	10615		5.8	3906	6555		5.3	5084.4	8795		13	12 408	12 555		213 46 720 £ 340 017

231

MATERIALS

The estimate in Tables 19.1 and 19.2 provides data for prototypes and two sequential production levels. Points to note on the spreadsheet are:

- Kitting, packing and shipping are included as direct labour costs.
- Pincount is included so that estimates of rework and visual inspection times can be added to the process operations cost sheet shown in Table 19.2.
- Yields on components are included. In this particular example of the costing approach, different yields for different components are shown. When this is done, the yield estimates can then be used by the purchasing department for ordering purposes. A simpler method is to apply an overall yield figure to the total cost.
- All individual component board area occupancies and heights are shown. These help in selecting and programming auto-placement machines.

19.7 Tooling costs

Subcontractors usually count themselves fortunate if, over the year, they receive payment from their customers to cover their real tooling costs. Average figures of 50 and 60 per cent are commonplace.

The lists of part-tooling charges given in the example quotations in Appendices 13A and 13B do not cover all the items that arise. Additional points would include:

- The cost of motorcycle couriers when the photomasters have to be remade because of redesign
- Pick-and-place machine tooling
- Adhesive deposition tooling and software program
- Special soldering operation jigs and fixtures
- Assembly jigs, e.g. for multi-board assemblies

A more complete list is given in Table 19.4.

20

Compound case-histories

A point is made that case-histories which detail nothing but the correct things that have happened in a particular successful project are never totally applicable to any other project.

The fact that most of us secretly enjoy being told about other people's mistakes is not a sufficient reason for writing the case-histories that follow. Mistakes cause companies to lose money, and surface mount technology has been fertile ground for this misfortune—particularly in the late 1980s. It has also cost people's jobs, and that is a good enough reason on its own for pointing up the pitfall areas.

Three potted case-histories are examined. They are blow-by-blow accounts of events over periods of six months or more. Although fictitious because they approach worst-case folly combinations, the individual mistakes are nevertheless real ones.

While they highlight the importance of training at all levels and/or the need for recruiting the right people before introducing a new assembly technology to production, these studies also emphasize the negative aspects of the current scene in electronics. Fortunately, for every failure there are more than ten success stories—though many of these may well have involved at least some of the financial scars of 'learning curve economics'.

The first history follows an ill-considered decision to manufacture in-house. The second covers some of the main problems that can arise when subcontracting in an amateur manner. The third shows the results of an SM 'greenhorn' OEM using a CAD bureau that had a non-professional management. Each case-history leads to a crisis that was avoidable.

20.1 Dataphlog Ltd: an in-house production disaster
Dataphlog Ltd, Basingtown, 1990

Dataphlog Ltd was a medium-sized original equipment manufacturer

(OEM) with a turnover of £20 million from 250 employees. It was developing and marketing portable data acquisition units suitable for use in monitoring and controlling warehouse inventories.

The company was applying conventional through-hole assembly methods on its own production lines, but, aware of the strong trend towards surface mounting, the development department had set up a small hand-assembly pilot line, allegedly 'to evaluate design and manufacturing issues' in SM technology. This unit was run part-time by an experienced development engineer and staffed by two skilled wiremen who were using hand placement and standard soldering irons to assemble mixed-technology boards.

Three models of a hand-held version of the company's previous product, but with enhanced functionality, had been put together—with difficulty. The component count was 300 (275 SM and 25 TH) on two boards measuring 60 mm × 150 mm. The technical director proudly laid them on the table at a Board meeting.

The Board members were delighted and almost mesmerized by the new toys. They decided to support the technical director (TD) and the advice that this form of miniaturization was essential for future products. A report from the TD on the most suitable technology options within the scope of SM was requested and the operations director was asked to prepare a budgetary estimate covering the equipment needed to produce 500 equipments per month. As an afterthought, the comparative costs of subcontracting all assembly work to a specialist SM company were also to be provided by the operations director.

The decision resulted in an immediate polarization of opinions within the company. Faction A wanted to set up an in-house manufacturing capability, a course of action favoured by the entire production department, which was keen to wipe out a history of indifferent product quality by winning new spurs.

Faction B favoured subcontracting—mainly to avoid the heavy capital investment required. This, needless to say, was the view of the finance director supported by the quality manager. The technical director sat firmly on the fence.

The managing director awaited the inputs the Board had demanded of his colleagues. A month later the technical department's advice emerged. Put briefly, it recommended that, to get the best combination of miniaturization, reliability and price, the company's future product range should be made using infra-red reflow-soldering of SM components on both sides of the printed circuit. The steadily decreasing percentage of through-hole (TH) components involved could be soldered in by hand.

Although devoid of direct experience in soldering SM components, the faction A plan, prepared by the production engineering department, recommended wave-soldering as the right solution because the company's existing machine could be used—at least for the first production phase.

A junior engineer had been sent to an exhibition to look for suitable processing plant, and relying on a supplier's statement that their brand of 'entry-level' automatic placement machine would handle 4000 components per hour, proposed the purchase of that equipment. The basic machine carried a price tag of £40 000, plus a similar sum for 80 separate component feeders. The same unit could also place metered amounts of adhesive from a syringe so that SM components could be glued to the underside of the board for their passage through the solder wave. A low-cost, unventilated static oven (price £500) could be purchased to cure the glue prior to soldering. Existing auto-insertion equipment would remain in use to handle the TH components.

The total capital cost was put at £100 000 and a sketch of the re-layout of the existing production area was provided. The entire area allotted to SM processes amounted to no more than 30 square metres.

The MD, an Oxford graduate in economic history, was delighted at the modest investment cost and decided to ask the Board to accept the faction A plan forthwith. Implementation of the technical director's recommendations would have involved a 25 per cent higher capital cost because of the need for a completely new soldering machine, and the quotations arriving from subcontractors showed a 10 per cent higher price than the projected internal prime cost calculated by the production manager. Everything looked clear-cut.

The members of faction B were at a disadvantage because the responsibility for preparing the subcontract plan was given to the production director. They suspected that enough optimism had been built into the in-house costing to ensure that it was lower than the lowest external bid. They decided to stay quiet for the time being as there was a rumour circulating that five years earlier Dataphlog's main competitor's circuit secrets were given away by their subcontractor to a third party: QED, subcontractors cannot be trusted.

Decision day for the directors duly arrived. Faction B remained silent because they had no evidence on which to base firm objections and, predictably, the Board decided to accept the 'make in-house' plan from faction A. However, there was one important proviso: the investment was to be divided into two equal tranches, with a review of progress before the second amount was allocated.

The first item to be specified and ordered from the UK supplier was the pick-and-place machine. Upon its delivery from abroad, complete with a foreign engineer in attendance to set it up, it was discovered that there was only one other similar unit in Europe and that all spares were retained at the manufacturer's factory. This was unfortunate, as the machine was found to be 'dead on arrival'.

A further minor difficulty was that it was not made clear at the outset by the UK agent that the existing Dataphlog factory compressed air supply was unsuitable. A highly filtered, oil-free supply and a reservoir local to the machine were needed. The unbudgeted cost turned out to be £6000.

The machine fault was diagnosed as a blown ASIC. One month later, a replacement printed board turned up by post. Meanwhile the foreign engineer had returned because his annual holiday was due. To avoid a repetition of this problem, the UK agent's English engineer had been sent overseas for training but would not be returning to fit the new ASIC until he had completed his course in three weeks' time.

To minimize further delay, an attempt was made to familiarize a Dataphlog technician with the machine using the operating manual. Although allegedly written in English, the words and grammar, while highly amusing in content, were not capable of unique interpretation. The attempt had to be abandoned and the manual was returned for editing.

When, eventually, the machine was ready for production trials, it was discovered that the placement rate quoted by the manufacturer was for small chip components on a 3 cm square printed board placed close to the pick-up head position. It did not contain any allowance for set-up time or for the reduced placement speed needed for ICs or for correcting feed and misplacement problems. Taking all non-operational times into account, the effective rate for a typical Dataphlog circuit content turned out to be less than half the originally projected figure. The machine's single-shift capability was therefore insufficient for the projected output requirement. In three months' time two-shift working would be essential and negotiations with the trade unions began immediately. This would add 6 per cent to the basic shop cost.

When the soldering trials using the company's existing wave-soldering machine were started, the joint defect rate was logged by the quality manager as approaching the disastrously high figure of 12 per cent. This means that the first-time post-soldering board assembly electrical test yield was zero. Immediately, six of the company's most experienced operators were sent on a training course to carry out rework on SM assemblies and an appropriate number of manual work-stations were ordered. These also were unbudgeted. Equipped with standard temperature-controlled soldering irons, they cost an extra £2000.

Meanwhile, frantic efforts were being made to discover the reason for the poor results. The solder bath was replenished with fresh solder but there was no improvement. The bath temperature was raised to 255 °C; again, there was no improvement.

A stronger flux was applied and the results were slightly better. Eventually the machine manufacturer was consulted and confirmed that, without an extended preheat section and a bolt-on vibrator for the bath, the existing equipment was not suitable for SM assembly. The modifications would take six weeks and cost £11 500.

The trained operators returned from their three-day course and advised their non-technical foreman that the standard soldering irons already ordered were unfit for the tasks and their use would prejudice the reliability of reworked components. Removing glued chips was best effected with heated tweezers so that rotation could be applied. Replacing or realigning chip ceramic capacitors would require hot gas pencils rather than soldering irons. The latter might be used for reflowing and lifting semiconductor lead joints one at a time, but for replacing such devices a multi-jet hot gas or a shuttered infra-red rework machine was essential. This would cost between £10 000 and £25 000, depending on the IC size coverage and degree of programmability required.

The wave-soldering machine modifications were completed, but results showed only a marginal improvement. A rising proportion of components were falling off the underside of the board into the bath. In high dudgeon, the MD of the machine supplier was summoned to Basingtown to explain why the changes had not provided the promised answer. Clearly, the modified machine was still 'unfit for purpose'.

The visitor asked to see the circuits concerned and realized immediately that the printed board layout was incorrect. The technical director was brought into the discussion and it became obvious that the designs put into pilot production were for reflow—not wave-soldering. Many components were far too close, and shadowing was preventing the wetting of many joints. A complete redesign of all boards was necessary before production using wave-soldering could become economic.

Production was stepped up to meet the planned ramp-up rate and to offset the low yields being achieved. Three more operators were trained to cope with the rising 'rework mountain' that was being generated.

An expert from the company supplying glue was summoned to explain why so many components were coming unstuck. He pointed out that, being unventilated, the oven would get full of vapours from the heated adhesive and these would inhibit its proper curing. For a few prototypes this effect might not be seen, but as the oven load was increased, the problem would grow. It would be necessary to purchase a small IR belt oven—price £3500, again unbudgeted.

The re-layout and procurement of new boards suitable for wave-soldering would take a minimum of six weeks and cost £2000. At the same time, a potential large customer from Canada had visited the factory and the technical director authorized some changes to the circuit design to improve its stability at low temperature. It was hoped that the extra components would arrive in time. They were three weeks late.

Eventually the modified circuits started to flow from the end of the production line and functional electrical testing of assembled boards began in earnest. The reject rate was below 15 per cent and everyone breathed a sigh of relief. The defective units were placed on separate shelves ready for fault diagnosis and rework.

A month later the MD toured the shop-floor and saw the growing shelf-loads of assemblies awaiting repair. What is wrong?

'We do not have enough technicians to probe and find out where the faults are and those that we have are doing the rework themselves. Often they have to remove good components to trace bad ones, but the main problem is to find the dry joints. Visual inspection is not efficient enough to do this, and anyway some of the joints are not visible because chip components have been located underneath or too close to large TH components. Often we do not have enough space to get the right rework tool in to do the job. Some of the technicians find that they cannot cope with the dexterity needed to make good SM solder joints.'

The technical director was again called to explain and, after consulting an expert friend, admitted that the importance of designing in-circuit test points for every component node on a 100 per cent basis and allowing enough space for visual inspection and rework had not been sufficiently understood. In any case, to do so would have meant adding 15 per cent to board area, and the cost of providing this type of test probe hardware and software would have increased the start-up costs by more than £3000.

The first shipments of 100 units was made to a key customer. Within a month, 27 were returned as non-functional. There was insufficient diagnostic effort available for rapid response, and the units were added to the still growing pile of defective units.

Redesign had been put in hand to allow in-circuit probe testing. Unfortunately, the existing plastic box was now not large enough and new moulding tools were needed. They would cost nearly £8000. Revised drawings were prepared in haste and sent both to the box manufacturer and to five key customers. 'Murphy's Law' operated, and the biggest and best customer replied that it was unable to accept the larger unit and cancelled its production order.

The 27 customer returns were at last diagnosed as having defective ceramic chip capacitors. The supplier was contacted in the expectation of giving its representative a carpeting with a threat that Dataphlog might have to go elsewhere for its supplies if capacitor quality didn't improve. All the faulty components, some of them in small fragments, were removed from the printed boards and put in a polythene bag ready for the confrontation.

On arrival, the representative, being technical, asked to see the production line and made discreet enquiries concerning the soldering processes. The non-technical foreman told him that the thermal profile seen by components had never been measured because it had been assumed that the modified wave-soldering machine was suitable for everything presented to it. The production engineer concerned confirmed that no measuring equipment had been ordered for checking such profiles—nobody had told him it was necessary.

The capacitor supplier arranged for a set of thermocouples and cables to be delivered on loan and kindly supervised their use. The profile turned out to be slightly on the high side but not really bad enough to cause the problem.

Regrettably, no measle charts had been kept by Dataphlog to identify repeated defects at the same point, but the technicians said that in their view the faults were random across the board. By chance, one of them identified the fact that several of the chip failures on one board were new capacitors replaced by him. Surely this must mean bad batches of replacement capacitors?

At last the cat was out of the bag! To speed up his work, the technician had been using one of the larger irons usually kept for TH component rework. Worse still, its bit temperature was measured at 385 °C. Excessive thermal shock will initiate minute internal cracks which would not be picked up at final test. Cyclic temperatures caused by daily switching the power on and off in the field would then bring propagation of the cracks, gradually leading to short-circuits between adjacent capacitor electrode layers or to open circuits.

Dataphlog's quality manager was unaware that it was important to identify the top surface of any chip capacitor before removing it for analysis of the cause of failure. This meant that, although the rogue soldering iron was probably the cause of the trouble, it would not now be possible to be certain because key evidence had been destroyed through ignorance. More experiments would be needed to make sure that all processes were satisfactory. This programme was put in hand at an estimated extra cost of £3500 in parts and labour.

The new in-circuit test rig arrived and immediately the effectiveness of the visual inspection and rework cycles improved. The first-time electrical yield of assembled and reworked circuits rose to 90 per cent, but this was still not good enough.

One of the customers had asked for the first 10 units due for shipment to be temperature-cycled to destruction to ensure that the product would be strong enough to meet its forecast lifetime in the specified environment. The test temperature range selected was between -10 °C and $+55$ °C and the estimated number of such cycles during life was 500. After 20 cycles all units were good. After 50 cycles there were two failures, and after 200 cycles all except one had failed. In nine cases out of ten, the failures were found to be due to open-circuit solder joints on a large ceramic chip capacitor.

At the same time, it was noticed that after the first rework/in-circuit test sequence, chip components and small transistors previously found good were becoming faulty. Initially rework operators were suspected of accidentally heating adjacent components when touching up solder joints, but the mystery deepened when close examination showed that this was not true.

What had changed? The new in-circuit test jig was not subjected to scrutiny until a visitor asked whether the bending of the board under the test probes was potentially harmful to some solder joints. The jig had been designed by an engineer used to TH assemblies who did not realize the damage that could be done to SM component joints by deformation of the printed board after assembly. No count had been made of the number of circuits that had passed through the jig before it was identified as the cause of failure. Some had been shipped, but of those which had not gone to customers it was necessary to assume that all completed sub-assemblies were suspect and to scrap or rework 150 units valued at nearly £50 000.

It was assumed (incorrectly, as it turned out) that the test jig must have been responsible for the above-mentioned temperature cycling failures and production was therefore allowed to proceed while further cycling tests were carried out. Morale slumped when the results were even worse than before, and the true cause was not discovered until it was far too late.

A further 250 units were shipped before the quality manager managed to persuade the MD to halt production. On the same day, two batches comprising 70 faulty circuits were returned by a customer.

At the next weekly progress meeting of executives, the finance director revealed the magnitude of the cash flow problem resulting from the unbudgeted investment costs and the effects of the projected customer-return scenario.

An emergency board meeting was convened and concluded that either the company must double its already overstretched borrowings to cover the hiatus, or a 'white knight' must be found. Meanwhile, a specialist consultant experienced in SM production technology was hired.

Over the next four weeks, product yield and quality improved slowly, but actual costs were still nearly 35 per cent higher than originally projected. A revised estimate of shop cost was prepared by the consultant. It indicated that, using the existing design and facilities, the original external quotes received by the operations director were unlikely to be beaten by in-house manufacture.

The poor temperature cycling results were attributed by the consultant to lack of knowledge in the electronic circuit and printed board layout teams. They had not understood the differential expansion cyclic stress phenomenon when large ceramic chip components are soldered directly to epoxy–fibreglass printed boards.

One month later the company was put in the hands of the receiver.
The Executive Summary of the consultant's report read as follows:
By failing to identify the need for training, first for himself and then for his staff, the MD must accept primary blame for the demise of the operation. At the same time, the Board was responsible for appointing a non-technical MD in a post which demanded more multi-discipline support than was available within the company.

The modest knowledge that existed in the development laboratory was ignored and no alternative source of help consulted. Apart from the error in layout, the wrong decision was taken on soldering method for the company's products. Soldering is the key process in surface mounting.

The successful application of SM technology is known to require full-time production staff who are strong on physics and chemistry. In a company dominated by electronics engineers, these skills may not be available. They are seldom present on the average electronic assembly shop-floor.

Similarly, the design team deploying surface mount assembly requires not only the above body of knowledge, but also far more mechanical engineering and stress analysis capability than in the past. The above stricture applies to the quality management team. In this case, faced with defects arising from phenomena beyong their ken, elementary mistakes were made and corrective action was either absent or far too late.

Finally, neither the Board nor the MD appear to have recognized the need for improvement in the intellectual strength of the company management when entering a higher technology level. Surface mounting is not regarded as being in the 'hi-tech' league, but the number of design and process variables is an order greater than for conventional TH assembly. For large boards holding more than a thousand components, it can even exceed those in semiconductor integrated circuit manufacture.

It is the successful management of these many and often conflicting variables that demands the extra training and mental capacity. Among smaller OEMs, the MD is very likely to be an entrepreneurial electronics engineer. Typically, he or she will have received little training in materials science or in applied physics, chemistry or mechanical engineering—all of which are as vital as circuit design knowledge in modern electronic equipment.

There are no substitutes for the hard core of multi-discipline technical knowledge needed for today's electronics 'make or buy' decisions. In the end, it boils down to the ability of the educational syllabus to keep pace with the rapidly changing needs of industry.

20.2 Buckpass Ltd: a subcontract tragedy of errors
Buckpass Ltd, Glenditch, 1989

Buckpass Ltd were manufacturers of computer hard disk units, including the associated drive control circuitry. They had a turnover of £45 million and a staff of 550.

The arrival of luggable and laptop portable computers had suddenly raised the market opportunity for small hard disk units containing all the required electronics packaged internally within the unit.

The skills developed by Buckpass in supplying robust units could be applied rapidly to the new miniaturized product range needed by computer manufacturers.

The Board had reviewed the production options for the first design. The chosen solution combined a largely proven electronic system with the first attempt at printed board layout using surface mount technology. Discussions were held with several experts in the latter field, including some with nearby potential subcontract assemblers who had served the company well in conventional assembly work but had minimal prior experience in surface mounting.

The high packing density in using mixed technology (combined SM and TH components on the same board) caused concern among the more experienced practitioners, especially in relation to manufacturing cost and reliability. The inexperienced subcontractors, seeing large orders on the horizon, proved less critical, and in the circumstances, were more likely to be believed.

The 'circumstances' referred to are the lack of knowledge of surface mount technology among Buckpass engineers and the responses of their project managers to those market forces pushing for ever smaller and lighter units. 'If the unit does not fit within the future space allowance conceived by our major customers, we will not have a product to sell.'

Because Buckpass was not yet a world-class company in the disk drive field, the idea of holding constructive discussions with the computer manufacturers and competitors on the best compromise between cost, size, shape and reliability was not followed up—largely through fear of upsetting powerful customers who take a 'know-all' stance, but also through Buckpass management's concern at the risk of exposing their new ideas to the competition.

The project go-ahead was given on the basis of setting up a small in-house surface mount pilot line to establish a firm body of knowledge before arranging subcontract assembly.

The first design contained no in-circuit test points. Three subcontractors, two of them local suppliers who had promised to set up an SM facility if an order was received and one a specialist SM house 400 miles away, were asked to quote for assembly and shipment after visual inspection only. No electrical test equipment was to be supplied to subcontractors and all items were to be 'free-issue' in kits for 500 or 1000 units. Buckpass would perform functional tests at their own goods inwards, and all fault diagnosis and rework was to be done by their in-house facility.

The chosen subcontractor had already made known its reservations concerning the manufacturability of the assembly, but when the circuit was received, no changes had been made and the deal was 'take it or leave it', with no price allowance for the extra difficulty and the added injunction from the buyer that the Buckpass in-house pilot production line was 'not having any problems with the assembly.'

The layout required surface mount and through-hole components on both sides of the board. It embodied the ideas of the more optimistic Buckpass engineers and their advisers in the inexperienced local subcontracting firms.

Contrary to expectations, the first order was awarded to the more distant and experienced supplier, and the first run produced 732 visually inspected circuits from the first kit of 1000 units, all of which were shipped to the customer four weeks later than promised. The initial visual inspection was carried out with a 100 per cent check on every solder joint, and the time taken for this was 49 minutes compared with the assembly labour time of 35 minutes. Rework took a further 24 minutes and inspection of the reworked items, 11 minutes. Of the 732 circuits received at the customer's goods inwards section, 214 were dead on arrival—mainly because of undetected dry joints—and a further 193 failed parametric tests because 10 kilo-ohm resistors had inadvertently been used in place of 100 kilo-ohm values.

Although the subcontractor had no formal product costing system in place at that time, it was estimated that their production cost was nearly three times higher than the forecast £31.00.

The in-house team spent an average of 1 hour 50 minutes per circuit searching for the dry joints and reworking them. In so doing they discovered a number of further faults not picked up at goods inwards and altogether added approximately £60 to the unit cost. Subsequently it emerged that the latter faults were components damaged during the subcontractor's rework operations. The buyer wasted no time in threatening non-payment unless all of the next batch of 1000 units were received with zero damage, i.e. 1000 units shipped, and a first-time electrical yield in excess of 90 per cent achieved.

The next 1000 kits arrived at the subcontractor's goods inwards and, after double checking, was found to have shortages on several key components. These were denied by the Buckpass chief buyer, who insisted that replacements should be paid for by the subcontractor. To protect further business, the latter reluctantly agreed.

The batch yielded 845 shippable units after the subcontractor had increased the visual inspection confidence level by double-checking. The amount of rework was made worse by poor solderability on three types of SOT 23 transistor, which had resulted in solder contaminating some of the gold-plated areas on the edge connector. The yield at the customer's goods inwards from the 845 units was improved to 81 per cent, still considerably below the customer's stated requirement.

A meeting between the technical directors of both companies was called to review the problems and consider remedial actions. The subcontractor tabled a long list of printed board layout design and component faults and also suggested that either the customer should perform an agreed goods inwards inspection schedule prior to shipping component and printed board kits, or he should pay for this to be done by the subcontractor. Both requests were turned down because of the claim that the in-house operation was not seeing anything like the same level of problems.

The subcontractor raised his price by 40 per cent and the OEM purchasing organization was ordered to bring in a second and third source, both of them also large, experienced surface mount specialist houses 400 miles away. Their quotes turned out to be very similar to the increased price demanded by the original subcontractor, but it was hoped that the yield and quality would be improved.

The first response from the new suppliers produced results very similar to those of the first subcontractor, but, in return for a follow-up order, one of them offered to attempt wave-soldering in place of the reflow method for which the circuit had originally been designed. This gave a slightly improved first-time yield, and the combined visual inspection and rework time was reduced to below 40 minutes—still quite unacceptable. The fact that reliability may have been impaired by the passing of critical components through the molten solder bath was not considered.

The managing director of one of the new subcontractors asked to visit the Buckpass in-house pilot production line and, after some hesitation, the request was granted. During his tour round the shop-floor he managed to extract statements admitting that the in-house results were far from satisfactory, with yields similar to those of the subcontractors. He was criticized for the amount of rework his product was causing and for deflecting the in-house efforts away from improving their own output. However, at the post-tour meeting summary it was agreed that considerable redesign was necessary, and the subcontractor MD offered to table his suggestions.

To the chagrin of the chief buyer, the design problems, the poor standard of component kitting and the lack of goods inwards inspection by Buckpass were used to negotiate a price increase at top management level.

The restrictions imposed by the existing case size prevented some of the most important changes from being implemented, but the remainder enabled significant improvements in the cost, and during the latter stages of

COMPOUND CASE-HISTORIES 245

the project a selling price of around £30 became practicable for the two additional subcontractors. The original assembler was not re-activated.

The sequence of events described above took place over a period of 20 months, during which time the Buckpass management had negotiated the setting up of their own assembly facility in the Far East. As this came on stream the indigenous supplier's outputs were progressively reduced to zero.

As with many other organizations before them, the search for lower costs by manufacturing in the Far East proved illusory. Collaboration between engineers proved expensive and unreliable, and the lesson was learned that low-cost labour cannot turn a poor design into a good product. Several of the fundamental manufacturing and reliability problems that had resulted from the initially defective packing density concept remained to haunt the product range. These, and the six months' time delay occasioned by them in the early stages, had effectively prevented the company from emerging as a world-class supplier. The well-established competitors gained more market shares and Buckpass was relegated to the sidelines.

A year later the company was drastically restructured in an effort to save it, but within 12 months the receiver was called in.

A summary of the mistakes leading to this unfortunate outcome is given below:

- Once the decision was made to subcontract SM assembly, the subcontractor should have been selected before printed board layout commenced. This would have enabled early feedback on achieving the best compromise between manufacturability, size and reliability and given the OEM the ability to minimize cost by tailoring the design to suit the subcontractor's assembly and soldering equipment.
- Design rules were non-existent and the advice of experienced practitioners was ignored in the struggle to minimize size. This contrasts greatly with the Japanese philosophy, in which the traduction of such rules is unthinkable and other ways of overcoming the problem are sought. In this case one obvious route was the increased use of integration via passive networks and ASICs to reduce the component count and board area occupancy.
- Management failed to appreciate the step-function in design knowledge needed by their engineers for successful SM product design, and, as in the first case-history, they were oblivious to the consequences of their failure to provide suitable training.
- No arrangements were made to carry out incoming inspection of boards and components. At commencement of the in-house facility installation, it is unlikely that sufficient knowledge of the pitfalls in SM component and board supply would have been available, and the task should have been given to the subcontractor *ab initio*.

- The attitude of the purchasing function was typical of that seen over the past 15 years when OEMs bully locally conventional 'board-stuffing' contractors. It is totally inappropriate for situations in which responsibility for mistakes is more likely to be with the OEM than with the subcontractor.

Here again, the importance of training the buyers is paramount.

20.3 Rufflay Ltd: a CAD bureau story
Rufflay Ltd, Strawtown, 1986

Rufflay, located in the south of England, have been one of the largest printed board design bureau houses for ten years. They employ a wide range of CAD equipment on several sites.

The advent of surface mount technology was seen as bringing new horizons to them, and in 1985 they announced themselves as experts and offered their customers a new service in printed board layout for surface mounted assemblies. Their reason for so doing was based on the acquisition of a commercial brochure from one of the major European component suppliers showing footprint pad layouts for the new surface mounted components. This, they felt, combined with their experience in conventional boards, was an excellent starting point. Without further inquiries, the information from the brochure had been fed into the library of one of their latest CAD machines, and they were all set to go.

Customers who, over the years, had used the company for conventional through-hole boards were delighted at the thought of receiving a helping hand from their long-time supplier. One such company was Cobblett, a nearby manufacturer of specialized medical apparatus for patient monitoring in private hospitals. Their need was to miniaturize a robust and heavy equipment which had to be transported from ward to ward on an unwieldy trolley and which, when connected to the patient, made access to the bedside somewhat restricted. The main body of the circuitry contained a large memory bank of more than 100 40-lead PLCCs assembled on three large printed boards and interconnected by a veritable bird's nest of cables, some of which were coaxial in order to handle the high-frequency parts of the system.

The instruction to Rufflay was to use their expertise to convert this jumble to a single large board using surface mounted components. On the promise of a one-month delivery for the layout, an order was placed in the sum of £1500. No alternative quotes were sought.

A bright young Rufflay draftsman, recently recruited to operate their latest CAD equipment purchased less than six months previously from the USA, was given the task. He set to work and within six weeks, using the machine's auto-router, came up with a single-sided eight-layer board measuring 26 in × 14 in.

Unaware that there was no pick-and-place machine on the market that could handle a board of that size, this was offered to the customer, who rejected it—not for that reason, but because it was 'too big to meet the miniaturization target'. Only then did Cobblett announce that they had been hoping for a board no larger than 10 inches square.

Back to the CAD machine and its auto-router, which, after a further four weeks' work, was able to offer a board with components on both faces, measuring 10 in × 12 in but having 16 layers, including four earth planes and four power planes.

In order to meet the small-size requirement, the auto-router had been programmed to work with a track width of 0.25 mm and all tracks on the board were of this width.

The customer was delighted and sought a local printed board manufacturer willing to have a go at the work. It was certainly the largest and most densely packed board he had ever seen, and in fact turned out to be only the third surface mount board he had ever attempted.

Upon examining the photomasters layer by layer, several problems were picked up:

- The layout was so densely packed that alignment tolerances required for the 16 layers meant that minimum clearances of 0.05 mm between vias and buried tracks would be frequent.
- It had been assumed that the resist apertures would be the same size as the footprint pads, and no clearance allowances had been made for inevitable misalignments between the resist and the pads.
- Vias had been designed on to the footprint pads and, owing to surface tension effects, would make dry joints inevitable at those points.

This information was fed back to Cobblett who, having already seen the programme drop 10 weeks behind schedule, were facing further delay and were understandably angry with Rufflay.

It was now impracticable for them to 'change horses in midstream', but while they awaited the third attempt at layout, Cobblett's chief engineer took limited details of the second version to a large and experienced surface mount assembly subcontractor for his comments. These came a week later on the same day as the third layout arrived. The bad news from Rufflay was that the board size had to be increased by 25 per cent to 15 in × 14 in to absorb the changes advised by the board manufacturer. The news from the experienced subcontract assembler was worse:

- There were no test points. No instruction had been given by Cobblett to

Rufflay to include them on a per-component basis; Cobblett were ignorant of the need—as indeed was the Rufflay draftsman.
- The type of resist to be used had not been specified to Rufflay. (This meant that the clearances allowed in the third design were much larger than necessary, as the board manufacturer had suggested those suitable for printed wet film resist on solder-coated copper. By using photo-imaged wet film on bare copper, the clearances around the pads could be halved, but this would mean going to a different board maker.)
- In order to achieve minimum board size, the spacing between the PLCC leads on adjacent packages was less than 1.0 mm. Rework would be extremely difficult and visual inspection of PLCC lead solder joints impossible.
- A surface mounted socket had been specified. This could bring reliability problems in its connections to the printed board.
- The experienced subcontractor was not prepared to accept the work unless all these points were corrected.

With the project running nearly three months late, the Cobblett board of directors met to consider what action to take. They decided to accept the risk of going ahead with the third Rufflay layout so that experience could be gained before a further attempt was made.

An order for supply of 50 boards was placed with the original local supplier and manufacture commenced. Concurrently, the purchasing manager was authorized to seek an alternative subcontract assembler having surface mount capability—preferably within a 50 mile radius.

As luck would have it, there turned out to be a small OEM close by who had recently set up an under-used in-house pilot line and was looking for work. He was given the components and awaited the boards—promised within two weeks.

Five weeks later, four boards arrived. It turned out that they were all that was left out of a batch of 60 started. Although the clearance problems had been addressed and cured, the auto-router was not of the 'intelligent' type and, perforce, had to be re-programmed for all tracks to be at 0.2 mm width throughout the board. For the particular problem concerned, this had led to severe yield problems in the manufacture of such a large board with so many layers and long fine tracks. This had led to a higher than acceptable level of track repairs—some buried and invisible.

The four boards were assembled by hand—with great difficulty—and shipped after a peremptory inspection of those solder joints that were visible and with no electrical test. The boards were tried in the Cobblett system and none of them worked.

Cobblett engineers set to work with soldering irons and after two days work decided that the same PLCC was incorrectly mounted on each board.

Unfortunately, it turned out to be a PLCC at the very heart of the array of 100 packages. The folly of the layout design had ensured that there was no room to get even the smallest soldering iron between the packages to melt the surface mount lead joints—let alone the multi-lead tip version which would melt them all at once. Thus it was necessary to remove all the outer PLCCs, one by one, to work a path in to access the offending package in the middle. Half way through the job it dawned on the team that it would be impossible to replace the packages singly and that they would have to go back to the subcontractor for a second pass through the reflow soldering machine.

The whole project was put on hold. By the time the rework had been completed, there was only one board left with all its surface copper tracking intact. This was tried and it worked.

A further batch of 10 boards was assembled. Six of them showed severe delamination after soldering and it was discovered that the boards had been left for several weeks in a damp, unheated store in the Cobblett building. All were scrapped. The remaining four were persuaded to function correctly.

A further batch of 10 boards that had also been in the same storage room were baked for 24 hours at 60 °C to dry them out. Even after this, five of them delaminated after soldering. An expert advised that the Rufflay draftsman had failed to appreciate the need to avoid large solid copper earthplane and powerplane areas on high-density multilayer surface mount boards and therefore did not opt for the cross-hatched pattern method of improving inter-layer adhesion.

At this point the board decided enough was enough. It cancelled the order on Rufflay and on the printed board manufacturer and postponed further work on surface mount designs until investigation and training programmes were completed. The costs written off amounted to nearly £200 000, and this excluded the potential loss of market share arising from the delays.

A summary of the key mistakes leading to this expensive learning curve event sequence is given below:

- The management of Rufflay must take the main share of blame for failing to realize that a set of component footprint pads did not constitute expertise in surface mount board layout. This error of judgement probably would have been nullified had the draftsman been given some training in the technology before being allowed to sit before the CAD system, i.e. a minimum of three months' working on a surface mount assembly line.
- The use of auto-routing engines for surface mount layouts is even now still in its infancy and many believe is suitable only for very simple small boards. The results on larger boards tend to be unsatisfactory as the engine will create extra layers rather than devise simpler alternatives.

- The necessary software intelligence to minimize the lengths of track taken below 0.3 mm wide wherever possible was not available and poor board production yields became inevitable. At that time, experienced surface mount board suppliers would probably have declined the order.
- It seemed that Cobblett management were unaware of the need to educate their own engineers in surface mount technology before embarking on the project. The result was too great a dependence on the Rufflay bureau and their board supplier. In particular, the need for their designers to know far more about surface mount printed board materials and manufacturing methods was not understood.
- CAD system software, unsatisfactory at the time of the project, is, in 1993, still far short of what is needed for good surface mount layout design.
- In fairness to Rufflay, it must be said that they were not alone in having brought mayhem to a customer's project. That particular lesson, and subsequent lessons learned by other OEMs who had similar problems, have taught that bureaux whose layout staff do not live daily with an operational surface mount assembly area are less likely to achieve good, manufacturable designs.

POSTSCRIPT

Unbeknown to senior management in either company, a fourth and final attempt at relayout was made in his own time by a conscience-stricken Rufflay engineer. This tried to overcome the main problem by using a new Japanese integrated circuit package containing the equivalent of eight of the PLCCs previously used. It was a quadpack with 160 leads, 40 on each side at 0.5 mm pitch—compared with the 10 leads per side at 0.05 inch pitch on the PLCCs.

The result was a layout which achieved the targeted 10 inch square board and also left adequate space for rework. Although this redesign was never realized in practice, a few prototype printed boards were made by backdoor methods and a sample batch of integrated circuits was received. By chance, a junior Cobblett engineer later discovered these and placed an integrated circuit on the board. It did not fit the pad layout.

Subsequent investigation revealed that the new Rufflay CAD system, although nominally gridless, was converting metric measurements into its basic Imperial format at only three decimal inch places, i.e. to the nearest 0.001 inch. The 0.5 mm lead spacing had been translated as 0.020 in (rounded from 0.019685 in), an error of 1.6 per cent. Multiplied by the 39 lead spacings on each side of the package, the cumulative error was 0.0126 in—more than half of one lead pitch.

By closing out the project when they did, the Cobblett management were wiser than they knew!

Warning: To cope with forthcoming 600-lead TAB devices with spacings at 0.1 mm, American inch-based CAD systems will require conversion capability to inches with at least seven decimal places.

Note
For managing directors and others who wish to know more about the technical issues leading to unacceptable yield losses and field failures referred to in the case-histories, the commonest of these are covered in Chapter 5 of the author's companion volume, *Surface Mount Guidelines for Process Control, Quality and Reliability* (McGraw-Hill, 1992), in the author's prior book, *Surface Mount and Mixed Technology PCB Design Guidelines* (Technical Reference Publications, Port Erin, Isle of Man, 1990), and also in Chapter 17 of this book.

21

Definitions and abbreviations

Note The definitions and abbreviations given below are my own. They apply to electronic assembly technology and associated business aspects and are solely for guidance in the context of this book.

As indicated in the Introduction, the use of the description 'printed circuit board' has been superseded in the IEC by the words 'printed board'. Throughout the text the new description has been stated in full, but for the purposes of this section only it has been abbreviated to 'PB'.

Acuity The ability of the eye to adjust its focus to objects at widely different distances.
AQL Acceptance quality level.
ASIC Application-specific integrated circuit.
Aspect ratio Ratio of length to width, e.g. of a PB.
ATE Automatic test equipment.

Bare board A printed board that has not been populated with components.
Break-out (from a PB) The process of breaking-out individual circuits from a step-and-repeat PB panel array. *See also* 'Step-and-repeat'.
Bridge Unwanted solder that shorts between adjacent conductors.

CAD Computer-aided design, e.g. of a PB layout.
CADMAT Computer-aided design, manufacture and test.
Catch 21.9 An *almost* impossible situation, as compared with Catch 22, a 'no-win' administrative muddle in which all solutions are defeated by mutually contradictory instructions.
CCTV Closed circuit television.
CERDIP Ceramic dual-in-line package (*see* DIL).
CFC Chloro-fluoro-carbon.

Chip-out A surface defect on a component or substrate in which a small chip of material has become detached from the main body.

Clamshell test rig A test rig in which both outer sides of an unpopulated substrate, e.g. a PB, and the connections between them can be electrically tested by a matrix of probes. The same type of rig is used for in-circuit testing of populated PBs.

Clinching The bending of through-hole component wires, tapes and tags after their insertion into a PB—to stop them falling out during handling and soldering operations.

COB 'Chip on board'. A generic term for an electronic assembly technology in which naked silicon chips are mounted directly on PBs and their connections to the printed wiring either wire-bonded or tape-bonded (*see* TAB) or by inversion in 'flip-chip' format.

CTE *See* TCE.

Cure/Curing The hardening of thermosetting resins and adhesives, often carried out at elevated temperature to speed up the process.

(D) Destructive. A description of a test.

DC Direct current.

DIL Dual-in-line. A package style for integrated circuits in which their outgoing leads appear in two parallel rows. Hermetic and moulded plastic versions exist.

ECIF Electronic Component Industry Federation. A UK trade association.

EDI/EDO Earliest date code in–Earliest date code out. A stores procedure.

EDP Electronic data processing.

Entec A commercial name for a flux laquer applied to printed board copper surfaces.

Fiducial mark An alignment mark on a PB capable of being recognized by an optical scanner, e.g. on an automatic pick-and-place machine to improve component placement accuracy.

FIFO First-in/first-out. A stores operating procedure.

Flexi-rigid (PB) A flexible PB incorporating local rigidized areas.

Flip chip A naked semiconductor die (chip) which can be mounted to a substrate by the inversion and joining of its outgoing connection pads via an intermediary material (e.g. solder) direct to correspondingly aligned pads on the substrate.

Free issues Parts supplied free of charge by an OEM to its assembly subcontractor.

FR4 A grade of fire-resistant printed board epoxy material.

Glass transition temperature (Tg) A concept describing the temperature at which an amorphous material (e.g. glass, plastic) starts to soften rapidly.

Hi-rel High reliability.
Hi-tech High technology.

IC Integrated circuit.
Ident Identification mark (e.g. for individual components on PBs).
IEC International Electro-technical Commission. The leading world standards authority for the electrical and electronics industries.
IEC(Q) Scheme An international scheme for qualifying factories which achieve set standards of quality management. BS 5750 is the UK part of the scheme. It is not intended for product approvals.
In-circuit test A method of probe testing individual and small groups of components after assembly. The same type of test rig can be used to check unpopulated PBs ('bare board testing').
Infra-red (IR) reflow soldering A process in which pre-positioned metered amounts of solder at component terminations and PB copper pad areas are heated to form joints by applying heat primarily in the form of infra-red light emission.

JEDEC Joint Electronic Device Engineering Council. A US standards organization defining standard component outlines.
JIT Just-in-time. A method of encouraging approved parts and materials suppliers to finance OEM inventories by holding stock and delivering it shortly (e.g. a few hours) before use on a production line.

Laser doppler vibrometry A method of detecting loose components by vibrating a populated printed board while it is illuminated with pulsed light, e.g. at 50 cycles/s.
Laser profile scanning A method of measuring solder joint contours and lead coplanarity by analysing reflected beams from a scanning laser system.
LCCC Leadless ceramic chip carrier. A cavity package style for integrated circuits in which the chip is mounted direct to a ceramic base with an 'hermetic' seal lid arrangement. Termination pads are sited on all four edges of the external side wall.
Leaching The dissolution of a metal coating into molten solder, e.g. from silver-based metallization on ceramic components.

Leadframe A pressed-out strip of individual metal component leads which are temporarily joined at their outer ends to enable connection to the component element in one movement. The part of the metal strip that joins them is cropped off after assembly.

Lead pitch The distance between the centre lines of adjacent leads emerging from the same face on a multi-lead component package.

LED Light-emitting diode.

MCM Multichip module.

MELF Metal electrode face component. Normally a leadless version of a cylindrical two-terminal conventional through-hole component (e.g. MELF diode, MELF capacitor, MELF resistor).

Mil American term for one-thousandth of an inch (0.001 in).

MIL STD Military Standard (US).

Mixed technology (MT) The combination of assembled surface mount and through-hole (conventional) components on the same PB.

'Mole' A proprietary name for a small portable temperature recorder capable of surviving passage through an infra-red soldering machine.

MTBF Mean time between failures.

(ND) Non-destructive. A description of a test.

Net list Network list. A list of all point-to-point connections within an electronic circuit.

OEM Original equipment manufacturer.

PB Printed board. (See Note at the beginning of this chapter.)

Photo-imageable resist A solder resist film which can be made insoluble by exposure to light. This property is used to define apertures in a resist layer on a PB by exposing it to ultra-violet light beneath an in-contact photographic positive (or negative).

Pick-and-place machine A high-speed programmable robotic machine which automatically picks out surface mount components from magazines, reeled tapes, chutes, etc., and places them in appropriate positions on a PB or other substrate.

PIH Pin-in-hole. Same as TH (see below).

PLCC Plastic, leaded chip carrier. A semiconductor integrated circuit package style.

'Popcorn effect' A type of semiconductor plastic package failure occurring during mass soldering.

p.p.m. Parts per million.

PTFE Polytetrafluorethylene. A plastic material.

PTH Plated-through hole. A hole in a PB which has been lined with metal by a plating process. Normally used to achieve inter-layer connections and/or to receive a through-hole component lead wire, tape or tag for soldering. *See also* 'Via'.

QED *Quod erat demonstrandum.* 'It has been demonstrated that . . .'

RA Rosin, activated. A classification of flux strength.
Reflow-soldering Creating solder joints by heating metered amounts of solder which have been preplaced, for example on a PB, using solder paste or solder preforms.
Resist/Solder resist A thin protective layer of plastic coating used to cover those areas of copper on the PB which are not required for component attachment or electrical test. *See also* 'Photo-imageable resist'.
RF Radio frequency.
RH Relative humidity.
RMA Rosin, mildly activated. A classification of flux strength.

Shrink-wrap A thin, transparent plastic packaging material that shrinks when heated and retains its shrunk condition upon cooling.
SIL Single-in-line. A package style in which all leads emerge in a single line from the package. SIL packages are often used for small hybrid (thick film or SM PB) circuits and resistor networks.
SIR Surface insulation resistance.
SM Surface mount/surface mounted.
SMA Surface mount assembly.
SMC Surface mount component.
SMD Surface mounted device (usually a semiconductor).
SMT Surface mount technology.
SO Small (previously Swiss) outline. A semiconductor integrated circuit surface mounted package style.
SOD Small-outline (surface mounted) diode.
SOIC Small-outline (surface mounted) integrated circuit.
SOJ Small-outline transistor, J-leads.
Solder paste A paste consisting of small solder particles suspended in a mixture of flux and rheological modifiers which enable deposition of precise amounts that retain their shape.
SOT Small-outline (surface mounted) transistor.
SPC Statistical process control.
Spec'n Specification.

Step-and-repeat An array of identical circuits (e.g. small PBs) joined together to form a large PB panel. *See also* 'Break-out'.

Submarining A condition in wave soldering when solder flows over the top side of a PB assembly—usually through routed slots or, if the board sags excessively, because it is too hot or too thin.

TAB Tape automated bonding. A method of attaching leads from passivated naked semiconductor chips to substrates, e.g. to a printed circuit board.

TCC Temperature coefficient of capacitance. The rate of change of capacitance with temperature, e.g. in p.p.m./°C.

TCE Temperature (thermal) coefficient of expansion. The rate of mechanical expansion of a material with the application of heat. Normally expressed in p.p.m/°C (sometimes CTE).

TCR Temperature (thermal) coefficient of resistance. The rate of change of resistance with temperature, e.g. expressed in p.p.m./°C.

Termination A conductive pad area on a leadless component, or a lead emanating from a component body, which is intended for electrical connection to a substrate.

Test Coupon A specially patterned region on an individual PB or on a panel of step-and-repeat circuits, used by a board manufacturer to check process performance and alignment accuracy.

TG *See* 'Glass transition temperature'.

TH Through-hole. A description of board assembly format in which component leads pass through holes in a printed circuit and are soldered therein.

Thermoplastic (material) A plastic material that can repeatably be softened by the application of the same level of heat, e.g. during and after moulding.

Thermosetting A process describing conversion of a mixture of plastic materials to hardened form by the application of heat over a defined period (curing). Once cured, thermosetting plastics normally withstand much higher temperatures than those used for curing.

Vapour phase reflow soldering A process in which pre-positioned metered amounts of solder paste at component terminations and PB copper pad areas are made to form joints by applying heat through the condensation of a non-oxidizing vapour. The vapour is created by heating a liquid whose boiling point is above the melting point of the solder used, e.g. 215–250 °C. The process is also called 'condensation soldering'.

Via A small plated-through hole in a PB which is used solely for interlayer connection.

Wave-soldering A soldering process in which joints are made *en masse* between pre-positioned component leads/terminations and PB copper pads, by passing them first through a flux applicator and then via a preheat zone through one or more waves of molten solder. In some cases the solder bath providing the wave either is agitated to assist penetration between closely spaced components, or acts as a source for a series of solder jets. In such machines, to remove bridges or other excess solder, these are necessarily followed either by a smooth wave or a hot air knife (jet).

WIP Work in progress. All product between pre-process stores and finished goods stores.

×4 Magnification by a factor of 4.

X-ray laminography Used to determine solder joint contour by comparing successive X-rays taken at different depths and reconstructing a three-dimensional image using a computer.

Yield A measure of manufacturing efficiency, e.g. the number of good circuits shipped from a batch divided by the number of kits supplied from stores for the batch, expressed as a percentage. The term is also applied to individual processes. The multiple of all the individual process yields gives the overall manufacturing yield.

Independent sources and bibliography

Independent sources of assistance

The Surface Mount Club
Building 15
National Physical Laboratory
Queens Road
Teddington
Middlesex, TW11 0LW

Tel (44) 81-943-7150
Fax (44) 81-943-6973
Telex 262344 G

Services offered
Management and technical consultancy
Selection and auditing subcontractors
Arbitration
Expert witness

ERA Technology Ltd
Cleeve Road
Leatherhead
Surrey, KT22 7SA

Tel (44) 372-374151
Fax (44) 372 374496
Telex 264046 G

Services offered
Contract research
Consultancy
Testing
Expert witness

The Institution of Electrical Engineers
Savoy Place
London, WC2R 0BL

Tel (44) 71-240-1871
Fax (44) 71-240-7735
Telex 261176 G

Services offered
President's list of consultants, expert witnesses, arbitrators

Books

Boswell, David, *Surface Mount and Mixed Technology, PCB Design Guidelines*, 1990. Technical Reference Publications, Port Erin, Isle of Man.

Boswell, David and Wickham, Martin W., *Surface Mount Guidelines for Process Control, Quality and Reliability*, 1992. McGraw-Hill Book Company Europe, Maidenhead, Berkshire, SL6 2QL.

British Standards Institution, Flatness of Electronic Assemblies, BSI document 91/248 29, IEC document 52 (United Kingdom) 217A.

Hinch, Stephen W., *Handbook of Surface Mount Technology*, 1988. Longman Group (UK), Harlow, Essex.

Holloman, James K. Jr, *Surface Mount Technology for PC Board Design*, 1989. Howard W. Sams & Co., 4300 West 62nd St, Indianapolis, IN46268, USA.

Lea Colin, *A Scientific Approach to Surface Mount Technology*, 1988. Electrochemical Publications, Ayr, Scotland.

Index

Note: Readers are advised that more than one reference may be present on any given page number.

Acceptance criteria, 126, 142, 149, 218
Acceptance quality levels (AQLs), 86, 219, 220
Access for:
   rework tools, 111, 114, 183
   visual inspection, 111, 113, 141
Acoustic inspection technique, 168
Added value, 23
Adhesion:
   of adhesives, 174, 194
   of PCB copper to base material, 185
   of soldered components to PCBs, 167
Adhesives,
   choice of, 184
   date coding of, 90
   goods inwards inspection, 90
   low free-ion count in, 160
   post-deposition checks, 113
   specifying, 90–1
   stock rotation, 90
   storage of, 90–1, 159–60
   use of, 91
Agreement terms, 28–9, 34–41
Alignment:
   in component placement, 163
   in PCB manufacture, 94
   of deposited adhesive, 163
   of printed solder paste, 162
Anti-static precautions:
   by subcontractors, 166, 174
   during assembly, 175–6
   during inspection, 176
   in component packaging, 87
   in rework, 176, 187
Applications:
   hybrid circuits, 60
   mixed technology assemblies, 58
   multichip modules, 68
   surface mounted assemblies, 57
Approval (factory), 172

Arbitration, 33
Artificial ageing, 89
Aspect ratio, 72, 113, 115
Auditing subcontractors, 64
   goods inwards inspection and stores, 159
   kitting and stock control, 160
   materials control, 158
   process control, 161
   purchasing function, 158
   quality function, 164
   visual standards, 167

Baking:
   printed boards, 105, 160
   semiconductor packages, 213
Bankruptcy, 73
Bare board testing, 101
Beam lead devices, 64
Bed of nails testing, 73
Binocular (stereo) microscopes, 167
Board accuracy expectations, 98
Bow in printed boards:
   affecting assembly processes, 72, 102
   affecting electrical test, 72, 117
   affecting reliability, 74, 100, 208
   specifying, 72, 100
Break-out, 208
Burn-in (*see* Screen testing)

CAD:
   bureaux, 19, 70
   costs, 224
   hardware, 19
   software, 19, 69, 130
   system, 69, 92, 115, 129
Calibration, 172
Cancellation, 38–40, 47
Capacity (manufacturing), 135
Capital investment for SM, 14–17
Cash flow, 7, 13, 30, 71, 154

INDEX

Ceramic chip capacitors, multilayer:
  chip-outs and cracks, 206, 209
  circuit design for, 208
  dielectric/electrode thickness, 206
  field failures, 207
  nickel barrier type, 206, 208
  rework of, 211
  specifying and controlling supply, 86
  testing and inspection, 87–8, 90, 209
CFCs (*see* Chlorofluorocarbons)
Changes:
  by the customer, 216
  by the subcontractor, 217
  controlling, 35, 143, 150, 165, 216
  notifiable, 217
Change note system, 165
Checking the net list, 115, 121
Chip body thickness, 208
Chip on board assembly
  advantages, 66
  assembly types, 64–5
  disadvantages, 66
Chip resistors:
  chip-outs and cracks, 89
  specifying and controlling supply, 86–8
Chlorine, 160
Chlorofluorocarbons (CFCs), 17
Choosing a customer, 152
Choosing a layout designer, 128
Circuit layout:
  design checks, 110
  design for SM manufacture, 110–1, 116
  for components, 111
  for electrical test, 114
  for reflow soldering, 112
  for reliability, 114
  for rework, 113, 207
  for visual inspection, 113
  for wave soldering, 111
Clamshell fixture, 101, 133
Cleaning:
  after rework, 188
  after soldering, 164
  before rework, 187
  hybrid circuits, 81
  pick and place machine jaws, 210
  print screens, 162
  printed boards, 101–2
  using ultrasonics, 71
  vacuum pencil suction pads, 182

Cling film, 102
Commercial security, 12, 42
Components:
  contamination on, 88
  co-planarity of leads, 86, 88
  damage to, 208
  dimensions of, 86, 88
  electrical testing, 85
  handling of, 18
  marking, 20
  misalignment (*see* Misalignment)
  placement of, 163
  post-soldering yields, 224, 227
  re-using, 182
  selecting for processes, 71
  selecting for SM circuits, 71–2
  solderability of, 4, 83
  storage life, 83
  weighing, 89
Condensation, 159
Conducting adhesives, 82
Confidentiality, 42
Connectors, 117
Contamination:
  on components, 87–8
  on hybrid circuits, 82
  on printed boards, 102
Contract types:
  full production, 27
  pilot production, 27
  prototypes, 26
Coplanarity of leads, 86, 88
Copper balance, 103
Copper thickness, 94
Cost–benefit analysis, 9
Costing:
  labour-based, 226
  machine-based, 226
  mixed labour/machine-based, 229
  rework, 224
  scarce resource, 226
Cost reduction, 33
Cost–size relationship, 73–4
Crossovers, 81
Customer aspects:
  design for manufacture and test, 157
  electronic design status, 71, 156
  financial status, 152–3
  marketing, 152–3
  quality, 156
  structural design status, 155
  technical, 154

Customer requirements, 118
Customer returns, 218, 126

Damage to components (*see* Components)
Date coding:
  adhesives, 90
  checking, 88–9
  components, general, 83, 147
  printed boards, 101, 104
  semiconductors, 83
  solder pastes, 90
Default, 47
Defects:
  classification of, 170–1
  in printed board panel arrays, 101
Deformation, mechanical:
  of chip components, 208
  of component leads, 90
  of printed boards, 102
Delamination:
  of multilayer ceramic capacitors, 89
  of printed boards, 102
Delivery, 23, 37, 43, 127, 142, 149
Departmental contacts, 215–6
Design:
  approval, 123
  authority, 125, 142, 150
  checking, 69
  error consequences, 148
  for component placement, 113
  rules, 18
  for electrical test, 111, 114, 157
  for manufacture, 157
  for reflow soldering, 112
  for reliability, 117, 141
  for rework, 113, 118
  for visual inspection, 113, 117–8
  for wave soldering, 111
  responsibilities, 125
  to meet customer/user requirements, 118
  validation, 129
Desiccants, 159–60
Designer knowledge, 3, 18, 69, 71, 128
De-soldering, 203
Destructive tests, 167
Differential expansion/contraction (*see* TCE mismatch)
Dimensioned drawings, 88, 93
Disputes, 36

Distributors, 19
Dowel pins and holes, 97, 132
Dry joints, 73, 183

Earliest date code in/out (EDI/EDO), 165
Electrical test:
  design for, 114
  equipment calibration, 172
  functional, 24
  hardware, 123
  in-circuit, 114, 169
  of components, 85, 89
  printed boards, probe systems for, 169
Electronic circuit design, 70, 79, 156
Encapsulation, 81
Enforcement, 42
Equipment structure, 74, 80, 115, 122, 155
Equipment lifecycle, 9
Exclusions, 44

Factory approval, 117, 123, 172
Fiducial (optical recognition) marks, 97
Field repair philosophy, 203
FIFO (see First in/first out)
Financial aspects:
  of selecting a subcontractor, 134
  of selecting a customer, 152
Financial risk spreading, 13
Fine pitch components, 114, 131–3, 191
Finished goods, 32
First in/first out (FIFO), 91, 159
Fitness for purpose, 118, 138, 158
Flatness of printed boards, 72, 117
Flooring, 176
Floppy disks, 93
Flux:
  low residue, 89
  no-clean, 89
  rosin active (RA), 218
  rosin mildly active (RMA), 83, 89, 201, 218
Fluxes, use of in:
  hot gas rework, 197
  infra-red rework, 199
  rework – general, 189
  thermode rework, 200–1
Focused infra-red beams (rework), 198
Force gauge, 173

Force Majeure, 47
Free issue components:
  inspection of, 108
  insurance of, 108, 126, 147
  managing, 23, 106–8
  responsibilities for, 123
  surplus, 147
  suitability of, 140
Functional electrical test, 73, 133, 219–20

Goods inwards inspection:
  components, 88, 159, 209
  chip resistors/capacitors, 88, 209
  dimensional checks, 88
  electrical checks, 85, 91
  free issue components, 108
  printed boards, 103
  semiconductors (SM), 88
  solderability, 89
  subcontracted assemblies, 142, 218–21
Gross margin, 5, 30
Guarantee, 37

Handling components, 174–6, 182
Harm occurrence ranking, 78
Hazard and risk analysis, 75
Health and safety, 46, 193
Heated electrode (*see* Thermode)
Heated electrode rework, 200
Heated tweezer rework, 194
Heat shrinking guns, 195
Heat sinking effects, 183
High-pincount devices, 72, 81, 197, 201
Hold harmless, 29
Hot gas pencils, 193
Hot gas rework machines, 195
Hotplates, 202
Housekeeping, 176
Hybrid circuits:
  advantages, 60
  applications, 60
  design checks, 79
  disadvantages, 61
  TCE mismatch, 82
  typical structures, 59

Identification marking:
  on components, 181
  on PCBs, 181

Illuminated magnifiers, 167
Image processing, 167
Inch-to-metric conversion, 130–1, 250–1
In-circuit electrical test, 73, 133, 155, 169
Indemnity, 29, 39, 222
Information technology, 3
Infra-red rework, 198
Inspection point chart, 50, 53
Inspection equipment:
  electrical, 169
  mechanical, 167
  thermal, 167
  visual, 167
Insurance, 126, 222
Intellectual property, 46
Interconnection technology, 3
Ionic contamination, 81, 102, 160

Jigging holes, 97
Jigs and tools, 125
Just in time (JIT), 12

Kitting, 160, 232

Large printed board assemblies, 73
Laser:
  as inspection tool, 88, 167
  for rework, 202
Layout rules/guidelines, 11, 71, 80, 140–1
Leaching:
  in rework, 191
  inspection of, 89
Learning curve, 225

Magnifiers, 167
Make or buy:
  cost–benefit analysis, 9
  key factors, 9, 11, 15
Manufacturer's drawings, 132
Market size:
  subcontract assembly, 3, 5
  world electronics, 3, 5
Materials:
  changes in, 158
  control of, 158
Mechanical testing, 167
MELF components: 72
  cost of, 71
  rework of, 195

264 INDEX

Metric-to-inch conversion, 130–1, 250–1
Microscopes, 167, 177–8
Miniature PLCCs (MPLCCs), 84
Misalignment:
  after component placement, 163
  after soldering, 163, 170–1
  in printed board manufacture, 94
  of adhesive, 163
Mismatch, TCE (*see* TCE mismatch)
Missing components, 170–1
Mixed technology assembly:
  advantages, 57
  applications, 58
  disadvantages, 58
  example structures, 58
Moisture penetration, 81, 212
Multichip modules: 66
  advantages, 67
  applications, 68
  disadvantages, 67
Multiple image processing, 167

Net list checking, 167
Nickel barrier, 206, 208
No-clean fluxes, 89
Nodes (circuit), 8, 143, 146, 157, 169
Non-destructive testing, 167–8
Notifiable changes, 217

Operating temperature, 123
Optical alignment marks, 97–8, 116, 132–3
Order acceptance, 26

Packing and shipping, 124, 228
Paint strippers (hot air), 195
Parts per million (p.p.m.):
  at customer goods inwards, 220
  post-soldering defects, 24, 73
Passive components:
  primary packaging, 88
  specifying, 86

Pastes (*see* Solder pastes)
Patents, 46
Payment terms, 38, 43, 144, 151
Peel strength, 184, 195
Photomasters, 92–3
Pick-and-place machine:
  component types handled, 136

Pick-and-place machine—*cont.*
  placement error, 99
  placement rate, 135
Plastic bags (*see* Polythene bags)
Plastic leaded chip carriers:
  PLCCs, 84, 88, 185
  MPLCCs (miniature), 84
Plating thickness, 84
Pneumatic dispensers, 202
Polythene bags, 87–8
Popcorn effect, 86, 160
Pre-bake (*see* Pre-heat)
Pre-baking assemblies prior to rework, 187
Pre-heat:
  for rework, 181, 188, 197
  for thermode soldering, 201
Price variations, 34
Primary packaging:
  cassettes, 88
  conductive materials, 176
  date codes on, 172
  of components, 87
  tapes, 87
  tubes, 87
  vials, 88
Printed board layout (*see* Circuit layout)
Printed board manufacturer:
  front-end system, 92
  selection of before design starts, 92
Printed board thickness, 93
Printed boards:
  accuracy expectations, 98
  aspect ratio, 72
  assembly process sequence, 122
  bare board testing, 101
  base material, 93
  bow (*see* Distortion)
  brushed tin-lead, 96
  cleanliness, 104
  conformance with net list, 103
  copper thickness, 94
  copper track accuracy, 94
  date coding, 101, 104
  defective circuits in panel, 101
  delamination (*see* Delamination)
  dimensional checks, 103
  dimensioned drawings, 93
  distortion, 100, 102
  electrical test, 101
  dowel holes, 103

Printed boards—*cont.*
  fiducial marks (*see* Optical marks)
  flatness, 72
  floppy disks, 93
  fluxed lacquer 94
  goods inwards inspection, 103
  hot gas levelling, 95
  hot liquid levelling, 95
  ident(ification) marking, 100
  intended assembly processes, 101
  issue numbers, 102
  jigging holes, 97
  large area, 73
  layout (*see* Circuit layout)
  long term storage, 105
  manufacturer, 92
  multilayering data, 94
  nickel-gold plating, 96
  non-soldered surfaces, 95
  optical marks, 97
  packing requirements, 102
  peel strength of copper, 185, 194, 195
  photomasters, 93
  plated-through holes, 99
  repair methods, 100
  solder resist material, 94, 184
  solderability, 104
  solderable surfaces, 95
  specification of, 71–2, 92, 121–2
  step and repeat patterns, 96
  storage of, 103, 105
  supplier's front-end system, 92
  test coupons, 101
  thickness, 93
  tin-lead plating, 95
  track defects, 100
  track repairs, 106
  transit packing, 102
  twist, 102
  vias in, 92, 99
  visual checks, 104
Priority control, 12
Process changes, control of, 165–7
Process control checks:
  after adhesive placement, 163
  after cleaning, 164
  after component placement, 163
  after in-circuit test, 164
  after reflow soldering, 163
  after solder paste deposition, 162
  after wave soldering, 163

Processes specified by the customer, 125
Product application fields, 120
Product file, 165
Product liability, 75, 221
Product life, 155
Product safety, 75, 118, 165, 221
Production control, 20, 161
Production yields, 94, 224
Pull strength, 167
Purchasing responsibilities:
  passive components, 86
  printed boards, 71, 75
  semiconductors, 85
Push testing, 167

Quad flatpacks (Quadpacks):
  dimensions, 88
  lead deformation, 88
Quality assurance testing, 126, 173
Quality function auditing, 164
Quality manual, 149, 166
Quality standards (SM assemblies), 149
Quantity variations, 32, 143, 150
Quotation examples:
  production prices, 43
  prototypes, 139
Quotations:
  basis for, 119
  defining the product, 119–20
  end product application, 120
  product function, 120
  technical provisos, 138, 139–41, 146–8

Realignment (*see* Tweaking)
Reel cover tape peel strength (*see* Peel strength)
Rejection of goods, 35, 218–20
Reliability, design for, 207
Representations, 44
Resist (*see* Solder resist)
Resistors (*see* Chip resistors)
Responsibilities, 215
  customer, 125
  subcontractor, 125
  purchasing department, 71, 75, 221
Rework:
  activity classification, 188
  addition of flux and solder, 189
  anti-static precautions, 187
  assessing equipment, 180

Rework—*cont.*
 chit, 169
 choice of adhesive for reflow soldering, 184
 cleaning prior to, 187
 component realignment (*see* Tweaking)
 component removal, 189
 component replacement, 191
 conventional soldering irons, 191, 202
 copper pad and track layout for, 184
 costing, 224
 design of layout for, 118, 182
 de-soldering tools, 203
 field repair philosophy, 204
 fluxes (*see* Fluxes in rework)
 focused infra-red light beams, 198
 handling components, 182
 heated tweezers, 194
 heat sinking effects, 183
 high-pincount devices, 197, 201
 hot gas pencils, 193
 hot gas rework machines, 195
 hotplates, 202
 laser de-soldering, 202
 masking of sensitive items, 188
 modified soldering irons, 195
 personnel and training for, 179
 pneumatic dispensers, 202
 pre-baking assemblies, 187
 pre-heating large boards, 188
 pre-heating sensitive components, 181
 printed board layout space constraints, 182
 printed board material type, 184
 realignment (*see* Tweaking)
 re-melting adjacent joints, 185, 197
 removal of adjacent components, 187
 removal of conformal coating, 187
 removal of solder, 191
 re-using components, 182
 RF-powered soldering irons, 192
 selecting equipment, 184–6
 thermal mass (*see* Thermal mass, in rework)
 thermal shock (*see* Thermal shock)
 thermode equipment, 200
 tweezers, 203
 unmarked components, 181

Rework—*cont.*
 vacuum pencils, 182, 203
 visual inspection after, 188
Risk, 30 (*see also* Title and risk)
Robin Hood tactics, 107
Rosin mildly active (RMA) flux (*see* Flux)
Routing slots, 96

Safety, 75, 165, 221
Safety profile:
 printed board assembly, 76
 television set, 77
Screen testing/burn-in, 124
Screen design, 87, 157, 174
Second sourcing, 23
Semiconductors, SM:
 device package integrity, 86
 electrical parameters, 85
 goods inwards inspection, 85, 88
 mechanical outline, 86
 primary packaging, 87
 specifying and controlling supply of, 85
Shear testing, 167
Shock testing, 167
SIR testing (*see* Surface insulation test)
Smoking, 176
SOD, 123, 195
Sodium, 160
Solderability:
 checking, 89
 component purchasing for good, 165
 discrete devices, 83
 goods inwards checks, 89
 integrated circuits, 83
 preserving printed board copper surfaces, 95, 102
 printed boards, 94, 105
 semiconductor leads, 83
Solderability testing, 89, 147
Solder coating thickness:
 on component leads, 83
 on printed boards, 95
Soldering equipment design, 7, 205, 207, 210
Soldering irons:
 miniature, 191
 modified, 195
 RF-powered, 192

Solder paste:
  general quality aspects, 91
  rejection of, 91
Solder resist:
  lifting of, 148
  material types, 94
  on printed boards, 94
  rework factors, 148
  specifying, 140, 194
SOT 23/143 packages, 84, 195
SOT 223 package, 84
SPC (*see* Statistical process control points)
Start-up losses, 21
Statistical process control points, 166, 169–71
Statutory regulations, 40
Stencil design, 87
Step and repeat, 101, 143, 145, 224
Stereo (binocular) microscopes, 167
Stereo projectors, 167
Stock control, 160–1
Stock evaluation, 31–2
Stock exhaust, 161
Storage:
  ambient conditions for, 123
  EDI/EDO procedures, 20, 160, 165
  inter-stage, 176
  of components, 83, 159
  of overmakes, 134
  of pastes and adhesives, 159–60
  of printed boards, 103, 160
  of print screens, 177
Stress analysis, 18, 70, 74
Structural options for board assemblies:
  mixed technology, 57–8
  surface mount technology, 52
  through hole technology, 48–9
Subcontractors, design:
  assessment of design capability, 128
  CAD system capability, 129
  design aspects, 130
  designer knowledge, 128
  job to be done, 130
Subcontractors, assembly and test:
  assessing production capacity, 135
  financial aspects, 134
  management of change, 165, 216–8
  production aspects, 135
  quality aspects, 156

Subcontractors, assembly and test—*cont.*
  stores and kitting aspects, 159–61
  team, 135
  types of, 6–8
Sub-subcontracting, 39
Sunlight, avoidance of, 163
Surface condition, 88
Surface insulation resistance (SIR) test, 102
Surface mount assembly, 52–6
  advantages, 56
  applications, 57
  disadvantages, 56
  example structures, 52
  process sequence, 53
Surface tension effects, 112

Tape automated bonding (TAB):
  advantages, 63
  applications, 133
  disadvantages, 64
  typical device structure, 62
Tape reels, 87
Tariff barrier, 2
TCE mismatch: 117
  fatigue failure due to, 75, 82, 208, 213
  structures to minimize effect of, 82, 214
Technical provisos in quotes, 138
Technologist support, 18
Technology stretch syndrome, 22
Temperature cycling (*see* Thermal cycling)
Tensile (pull) testing, 167
Termination of contract, 38–40, 47
Terms and Conditions of Sale, 26, 28, 42, 127
Terms and Conditions of Purchase, 26, 34, 127
Test equipment calibration, 172
Test houses, 7
Thermal cycling:
  design integrity validation, 124
  in soldering, 86
  of joints by usage, 75, 114, 117
  tests, 168, 173, 209
  to minimize the effects of, 82, 209
Thermal mass:
  copper on printed boards, 113, 183
  in rework, 183

Thermal mass—*cont.*
  laser I–R test, 167
Thermal profile, 210
Thermal shock:
  chip capacitors, 206
  component cracking due to, 206
  during assembly, 208
  during rework, 181, 183, 188, 190, 195
  popcorn effect, 86, 160
  pre-heating components before rework, 181
  pre-heating printed boards, 188
  semiconductors, 86, 160
  soldering equipment design, 78, 205, 207
Thermode soldering:
  description of process, 200
  rework, 200–1
Through-hole assembly:
  advantages, 49
  disadvantages, 51
  example structures, 49
  process sequence, 50
Title and risk, 36, 45
Tooling charges, 139, 145, 232
Toxic fumes, 187
Traceability, 37, 166, 172, 174
Track defects, 100
Track repairs, 100
Training:
  board layout engineers, 11, 18–19
  log books, 166
  operators, 11, 20–1
  programmes, 165
  top management, 10, 14
Trouble-shooting, 177

Turnkey projects, 30
TV cameras, 167
Tweaking, 188
Tweezers, 194
Type approval testing, 25, 55, 141–2, 209

Ultrasonics (*see* Cleaning)

Vacuum pencil tips, 182
Vertical integration, 11
Vias:
  electrical test points, 114
  visual checks on, 178
Vials, 88
Vibration testing, 167
Visual inspection:
  after rework, 203
  design for, 118
  equipment for, 167
  methods, 133, 167
  standards, 156, 167

Warranties, 44
Wave soldering:
  printed board layout for, 111–2
  pre-heat, 210
Weighing components, 88
Work in progress (WIP), 32

X-ray:
  inspection, 167
  laminography, 167

Yields (*see* Production yields)

Zero defects, 166